SpringerBriefs in Physics

For further volumes:
http://www.springer.com/series/8902

Francesco Sannino

Dynamical Stabilization of the Fermi Scale

Towards a Composite Universe

 Springer

Francesco Sannino
CP3-Origins, Center for Particle Physics
University of Southern Denmark
Odense M
Denmark

ISSN 2191-5423   ISSN 2191-5431   (electronic)
ISBN 978-3-642-33340-8   ISBN 978-3-642-33341-5   (eBook)
DOI 10.1007/978-3-642-33341-5
Springer Heidelberg New York Dordrecht London

Library of Congress Control Number: 2012948634

Printed on acid-free paper

Springer is part of Springer Science+Business Media (www.springer.com)

To Lene, Maria Sofia and
Matthias Alessandro

Preface

Strong dynamics constitutes one of the pillars of the standard model of particle interactions, and it accounts for the bulk of the visible matter in the universe. It is therefore a well-posed question to ask if the rest of the universe can be described in terms of new highly natural four-dimensional strongly coupled theories. The goal is to provide a coherent overview of how new strong dynamics can be employed to address the relevant challenges in particle physics and cosmology from composite Higgs dynamics to dark matter and inflation. We will first introduce the topic of dynamical breaking of the electroweak symmetry also known as technicolor. The knowledge of the phase diagram of strongly coupled theories plays a fundamental role when trying to construct viable extensions of the standard model. Therefore, we present the state-of-the-art of the phase diagram for gauge theories as function of the number of colors, flavors, matter representation, and gauge group. Recent extensions of the standard model featuring minimal technicolor theories are then introduced as relevant examples. We finally show how technicolor or in general new strongly coupled theories can lead to natural candidates of composite dark matter and inflation.

Odense, June 2012   Francesco Sannino

Contents

Chapter 1
Technicolor Prelude

Abstract The energy scale at which the Large Hadron Collider experiment (LHC) operates is determined by the need to complete the standard model of particle interactions. In particular LHC is set to unveil the origin of mass of any known elementary particle, i.e. the Higgs mechanism. Together with classical general relativity the standard model constitutes one of the most successful models of nature. We shall, however, argue that experimental results and theoretical arguments call for a more fundamental description of nature. Technicolor offers a physical mechanism underlying the Higgs sector of the standard model. We will review strengths and weaknesses of technicolor calling for novel type of strong dynamics.

1.1 The Need to go Beyond

In the first figure we schematically represent, in green, the known forces of nature. The standard model of particle physics describes the strong, weak and electromagnetic forces. The yellow region represents the energy scale around the TeV scale and being explored directly at the LHC, while the red part of the diagram is speculative Fig. 1.1.

All of the known elementary particles constituting the standard model fit on the postage stamp shown in Fig. 1.2. Interactions among quarks and leptons are carried by gauge bosons. Massless gluons mediate the strong force among quarks while the massive gauge bosons, i.e. the Z and W, mediate the weak force and interact with both quarks and leptons. Finally, the massless photon, the quantum of light, interacts with all of the electrically charged particles. The standard model Higgs does not feel, directly, strong interactions. The interactions emerge naturally by invoking a gauge principle. It is intimately linked with the underlying symmetries relating the various particles of the standard model.

The asterisk on the Higgs boson in the postage stamp indicates that the discovery of a new state with properties similar to the standard model Higgs has been announced on the 4th of July 2012 by the ATLAS and CMS experimental collaborations at CERN. More details and physical implications can be found in the next section.

F. Sannino, *Dynamical Stabilization of the Fermi Scale*, SpringerBriefs in Physics,   1
DOI: 10.1007/978-3-642-33341-5_1, © The Author(s) 2013

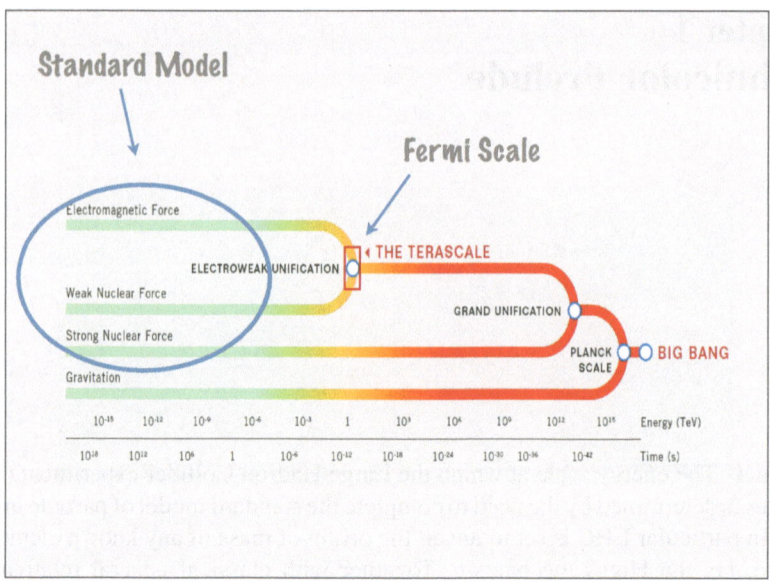

Fig. 1.1 Cartoon representing the various forces of nature. At very high energies one may imagine that all the low-energy forces unify in a single force

Fig. 1.2 Postage stamp representing all of the elementary particles which constitute the standard model. The forces are mandated with the $SU(3) \times SU(2) \times U(1)$ gauge group

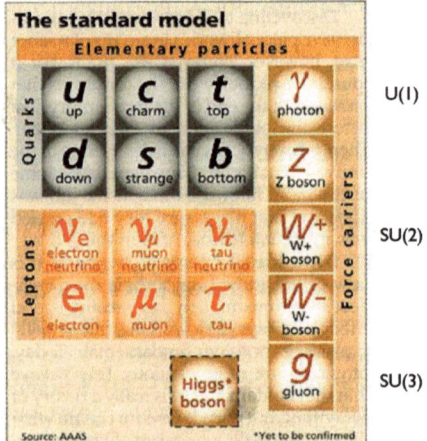

Intriguingly the standard model Higgs is the only fundamental scalar of the standard model.

The standard model can be viewed as a low-energy effective theory valid up to an energy scale Λ, as schematically represented in Fig. 1.3. Above this scale new interactions, symmetries, extra dimensional worlds or any other extension could emerge. At sufficiently low energies with respect to this scale one expresses the existence of new physics via effective operators. The success of the standard model

Fig. 1.3 The standard model can be viewed as a low-energy theory valid up to a high energy scale Λ

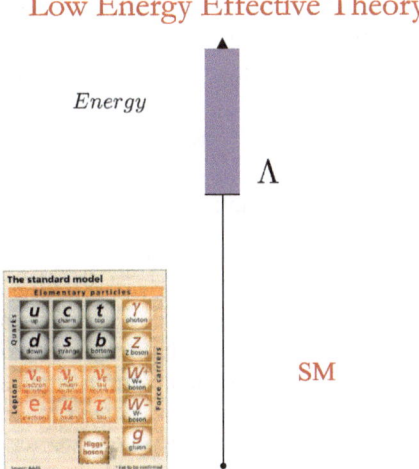

Low Energy Effective Theory

Energy

Λ

SM

is due to the fact that most of the corrections to its physical observables depend only logarithmically on this scale Λ. In fact, in the standard model there exists only one operator which acquires corrections quadratic in Λ. This is the squared mass operator of the Higgs boson. Since Λ is expected to be the highest possible scale, in four dimensions the Planck scale, it is hard to explain *naturally* why the mass of the Higgs is of the order of the electroweak scale. This is the hierarchy problem. Due to the occurrence of quadratic corrections in the cutoff this standard model sector is most sensitive to the existence of new physics.

1.2 The Higgs and its Scent

It is a fact that the standard model Higgs allows for a direct and economical way of spontaneously breaking the electroweak symmetry. It generates simultaneously the masses of the quarks and leptons without introducing flavor changing neutral currents at the tree level. The Higgs sector of the standard model possesses, when the gauge couplings are switched off, an $SU_L(2) \times SU_R(2)$ symmetry. The full symmetry group can be made explicit when re-writing the Higgs doublet field

$$H = \frac{1}{\sqrt{2}} \begin{pmatrix} \pi_2 + i\pi_1 \\ \sigma - i\pi_3 \end{pmatrix} \tag{1.1}$$

as the right column of the following two by two matrix:

$$\frac{1}{\sqrt{2}} (\sigma + i\tau \cdot \pi) \equiv M. \tag{1.2}$$

The first column can be identified with the column vector $\tau_2 H^*$ while the second with H and τ^2 is the second Pauli matrix. The $SU_L(2) \times SU_R(2)$ group acts linearly on M according to:

$$M \to g_L M g_R^\dagger \quad \text{and} \quad g_{L/R} \in SU_{L/R}(2). \tag{1.3}$$

One can verify that:

$$M \frac{(1 - \tau^3)}{2} = [0, H]. \quad M \frac{(1 + \tau^3)}{2} = [i\tau_2 H^*, 0]. \tag{1.4}$$

With $[0, H]$ we mean that this is a two by two matrix with the first column made by zeros and the second column is made by the H entries. Similarly for $[i\tau_2 H^*, 0]$. The $SU_L(2)$ symmetry is gauged by introducing the weak gauge bosons W^a with $a = 1, 2, 3$. The hypercharge generator is taken to be the third generator of $SU_R(2)$. The ordinary covariant derivative acting on the Higgs, in the present notation, is:

$$D_\mu M = \partial_\mu M - ig W_\mu M + ig' M B_\mu, \quad \text{with} \quad W_\mu = W_\mu^a \frac{\tau^a}{2}, \quad B_\mu = B_\mu \frac{\tau^3}{2}. \tag{1.5}$$

The Higgs Lagrangian is

$$\mathscr{L} = \frac{1}{2} \text{Tr} \left[D_\mu M^\dagger D^\mu M \right] - \frac{m^2}{2} \text{Tr} \left[M^\dagger M \right] - \frac{\lambda}{4} \text{Tr} \left[M^\dagger M \right]^2. \tag{1.6}$$

At this point one *assumes* that the mass squared of the Higgs field is negative and this leads to the electroweak symmetry breaking. Except for the Higgs mass term the other standard model operators have dimensionless couplings meaning that the natural scale for the standard model is encoded in the Higgs mass.[1]

At the tree level, when taking m^2 negative and the self-coupling λ positive, one determines:

$$\langle \sigma \rangle^2 \equiv v_{weak}^2 = \frac{|m^2|}{\lambda} \quad \text{and} \quad \sigma = v_{weak} + h, \tag{1.7}$$

where h is the Higgs field. The global symmetry breaks to its diagonal subgroup:

$$SU_L(2) \times SU_R(2) \to SU_V(2). \tag{1.8}$$

To be more precise the $SU_R(2)$ symmetry is already broken explicitly by our choice of gauging only an $U_Y(1)$ subgroup of it and hence the actual symmetry breaking pattern is:

---

[1] The mass of the proton is due mainly to strong interactions, however its value cannot be determined within QCD since the associated renormalization group invariant scale must be fixed to an hadronic observable.

$$SU_L(2) \times U_Y(1) \rightarrow U_Q(1), \tag{1.9}$$

with $U_Q(1)$ the electromagnetic abelian gauge symmetry. According to the Nambu-Goldstone's theorem three massless degrees of freedom appear, i.e. $\pi^\pm$ and π^0. In the unitary gauge these Goldstones become the longitudinal degree of freedom of the massive elecetroweak gauge-bosons. Substituting the vacuum value for σ in the Higgs Lagrangian the gauge-bosons quadratic terms read:

$$\frac{v_{weak}^2}{8} \left[g^2 \left(W_\mu^1 W^{\mu,1} + W_\mu^2 W^{\mu,2} \right) + \left(g W_\mu^3 - g' B_\mu \right)^2 \right]. \tag{1.10}$$

The Z_μ and the photon A_μ gauge bosons are:

$$Z_\mu = \cos \theta_W W_\mu^3 - \sin \theta_W B_\mu,$$
$$A_\mu = \cos \theta_W B_\mu + \sin \theta_W W_\mu^3, \tag{1.11}$$

with $\tan \theta_W = g'/g$ while the charged massive vector bosons are $W_\mu^\pm = (W^1 \pm iW_\mu^2)/\sqrt{2}$. The bosons masses $M_W^2 = g^2 v_{weak}^2/4$ due to the custodial symmetry satisfy the tree level relation $M_Z^2 = M_W^2/\cos^2 \theta_W$. Holding fixed the EW scale v_{weak} the mass squared of the Higgs boson is $2\lambda v_{weak}^2$ and hence it increases with λ. We recall that the Higgs Lagrangian has a familiar form since it is identical to the linear σ Lagrangian which was introduced long ago to describe chiral symmetry breaking in QCD with two light flavors. We will discuss this formal similarity in the next sections.

Besides breaking the electroweak symmetry dynamically the ordinary Higgs serves also the purpose to provide mass to all of the standard model particles via the Yukawa terms of the type:

$$- Y_d^{ij} \bar{Q}_L^i H d_R^j - Y_u^{ij} \bar{Q}_L^i (i\tau_2 H^*) u_R^j + \text{h.c.}, \tag{1.12}$$

where Y_q is the Yukawa coupling constant, Q_L is the left-handed Dirac spinor of quarks, H the Higgs doublet and q the right-handed Weyl spinor for the quark and i, j the flavor indices. The $SU_L(2)$ weak and spinor indices are suppressed.

When considering quantum corrections the Higgs mass acquires large quantum corrections proportional to the scale of the cutoff squared.

$$M_{H_{\text{ren}}}^2 - M_H^2 = \frac{kg^2 \Lambda^2}{16\pi^2}. \tag{1.13}$$

Here g is and electroweak constant and k a numerical factor depending on the specific model, expected to be $\mathcal{O}(1)$. Λ is the highest energy above which the standard model is no longer a valid description of nature and a large fine tuning of the parameters of the Lagrangian is needed to offset the effects of the cutoff. This large fine tuning

Fig. 1.4 Values of the Higgs
mass from the standard fit
(which does not take into
account direct Higgs searches)
obtained by excluding different electroweak observables.
The *green* band represent the
1σ error range around the best
fit value of M_H

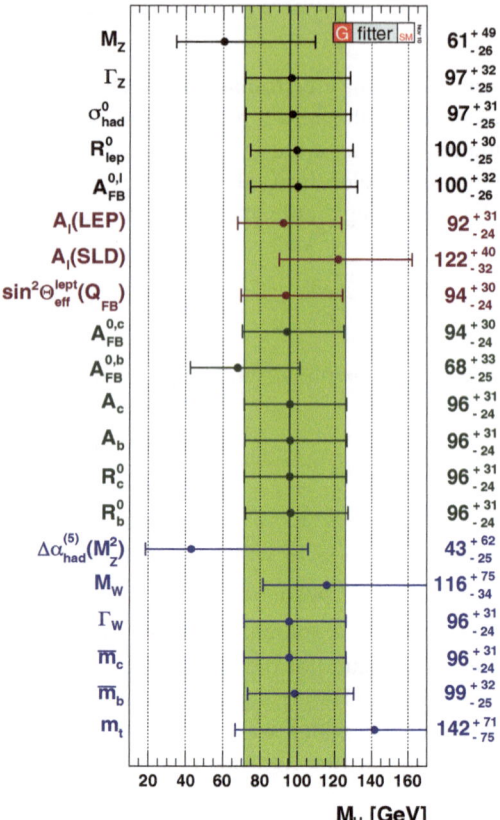

is needed because there are no symmetries protecting the Higgs mass operator from
large corrections which would hence destabilize the Fermi scale (i.e. the electroweak
scale). This problem is the one we referred above as the hierarchy problem of the
standard model Fig. 1.4.

The constant value of the Higgs field evaluated on the ground state is determined
by the measured mass of the W boson. On the other hand, the value of the standard
model Higgs mass (M_H) is constrained only indirectly by the electroweak precision
data. The preferred value of the Higgs mass (obtained by the standard fit which
excludes direct Higgs searches at LEP and Tevatron) is $M_H = 95.7^{+30.6}_{-24.2}$ GeV at
68 % confidence level (CL) with a 95 % CL upper limit $M_H < 171.5$ GeV, as given
by the generic fitting package Gfitter [1]. The corresponding results obtained by
a fit including the direct Higgs searches produces $M_H = 120.6^{+17.9}_{-5.2}$ GeV at 68 %
confidence level (CL) with a 95 % CL upper limit $M_H < 155.3$ GeV, as reported on
http://gfitter.desy.de/GSM/ by the Gfitter Group.[2]

---

[2] All the plots and numerical results we use in this section are reported by the Gfitter Group and
can be found at the web-address: http://gfitter.desy.de/GSM/.

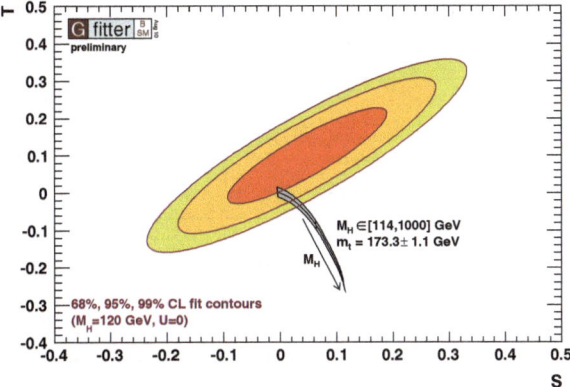

Fig. 1.5 The 68, 95, and 99 % CL contours of the electroweak parameters S and T determined from different observables derived from a fit to the electroweak precision data. The *gray* area gives the standard model prediction with m_t and M_H varied as shown. $M_H = 120$ GeV and $m_t = 173.1$ GeV defines the reference point at which all oblique parameters vanish

The final result of the average of all of the measures, however, has a Pearson's chi-square (χ^2) test of 17.5 for 14 degrees of freedom. A Higgs heavier than 155.3 GeV is compatible with precision tests if we allow simultaneously new physics to compensate for the effects of the heavier value of the mass. The precision measurements of direct interest for the Higgs sector are often reported using the S and T parameters as shown in Fig. 1.5. From this graph one deduces that a heavy Higgs is compatible with data at the expense of a large value of the T parameter. Actually, even the lower direct experimental limit on the Higgs mass can be evaded with suitable extensions of the standard model Higgs sector.

Direct searches results, updated July 2012, from the Large Hadron Collider (LHC) experimental collaborations exclude the standard model Higgs mass in the following mass ranges 111.7–121.8 and 130.7–523 GeV at the 99 % confidence level [2, 3] in agreement with the latest Fermilab Tevatron results [4]. See also for earlier published results [5–10] while the combined LEP2 results exclude it below 114.5 GeV at the 95 % confidence level [11].

Excitingly in the low mass window for the Higgs, not excluded by experiments, both ATLAS and CMS collaborations at LHC have independently reported the discovery[3] of a new particle [2, 3] with properties close to the standard model Higgs. In Fig. 1.6 are shown the ATLAS results, updated to July 2012, for the two relevant decay channels of the Higgs which have driven the discovery of the new particle.

The burning question is clearly whether the new particle state is indeed the standard model Higgs. In fact, given the current experimental status it is not yet possible to establish with certainty that the newly established state is the missing Higgs boson

---

[3] Corresponding to, at least, a 5σ deviation from the background.

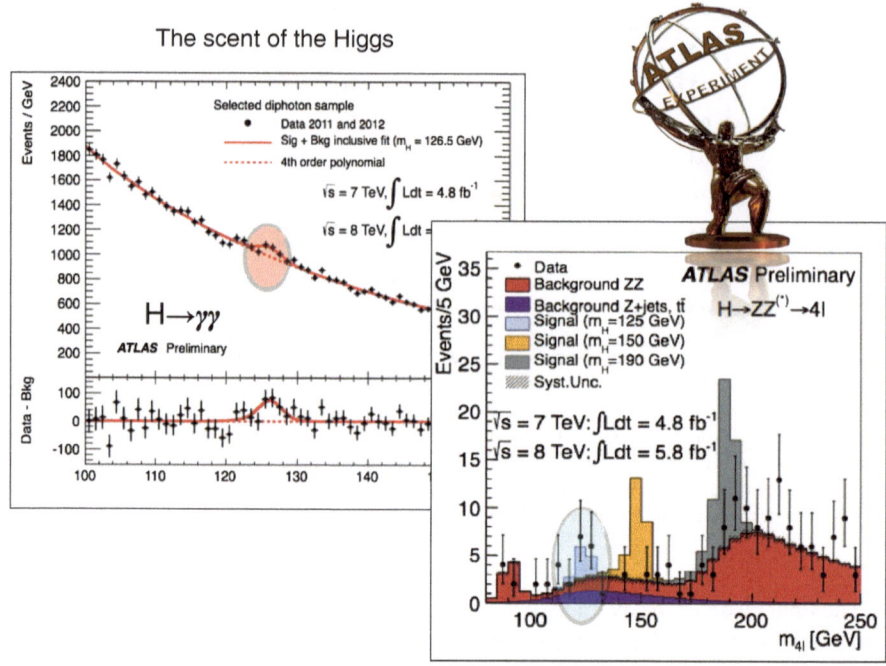

Fig. 1.6 ATLAS results [2] for the two processes associated to the searches for Higgs decaying in two photons (*upper left*) and four leptons (*bottom right*). Both plots corresponds to the associated number of events as function of the final state invariant mass measured in GeV. The *shaded ellipses* indicate the correlated excesses in the two processes allowing to determine the mass of the new particle state. Similar results have been reported by the CMS collaboration [3]

although it does smell like it. To determine the nature of the new state the experiments are studying how it is produced and how it decays into standard model particles.

As for the implications of this discovery on our understanding of Nature we can already say that albeit the nature of the new particle is not fully determined its presence is related to the puzzle of the origin of mass of every elementary particle. One can envision, at least, two major logical possibilities: (i) The new state is a fundamental boson as predicted within the Higgs sector of the standard model or its supersymmetric extensions. By fundamental here we mean that it is not made out of something else. If this were confirmed it would be important given that no other elementary spin zero boson has ever been discovered in Nature. (ii) The other possibility is that the new state is not elementary but composed of more fundamental objects. As the proton and neutron are composed by quarks likewise this state may be composed by new type of quarks. In this scenario experiments at CERN have the chance to discover many more new composite particles built from rearranging the new type of quarks in different combinations. More exotic possibilities have been envisioned by theorists and are not yet excluded by the current experimental results.

Therefore the new discovers begs the questions: Is the new state composite? How many Higgs fields are there in nature? Are there hidden sectors? Is the standard model written in magnetic or electric variables [12]?

1.3 Riddles

Why do we expect that there is new physics awaiting to be discovered? Of course the discovery of the presumed Higgs is an extraordinary feat for our understanding of Nature but leaves us with a standard model which has both conceptual problems and phenomenological shortcomings. In fact, theoretical arguments indicate that the standard model cannot constitute the ultimate description of nature:

- **Hierarchy Problem**: The Higgs sector is highly fine-tuned. We have no natural separation between the Planck and the electroweak scale.
- **Strong CP Problem**: There is no natural explanation for the smallness of the electric dipole moment of the neutron within the standard model. This problem is also known as the strong CP problem.
- **Origin of Patterns**: The standard model can fit, but cannot explain the number of matter generations and their mass texture.
- **Unification of the Forces**: Why do we have so many different interactions? It is appealing to imagine that the standard model forces could unify into a single Grand Unified Theory (GUT). We could imagine that at very high energy scales gravity also becomes part of a unified description of nature.

There is no doubt that the standard model is incomplete since we cannot even account for a number of basic observations:

- **Neutrino Physics**: Only recently it has been possible to have some definite answers about properties of neutrinos. We now know that they have a tiny mass, which can be naturally accommodated in extensions of the standard model, featuring for example a see-saw mechanism. We do not yet know if the neutrinos have a Dirac or a Majorana nature.
- **Origin of Bright and Dark Matter**: Leptons, quarks and the gauge bosons mediating the weak interactions possess a rest mass. Within the standard model this mass can be accounted for by the Higgs mechanism, which constitutes the electroweak symmetry breaking sector of the standard model. However, the associated Higgs particle has not yet been discovered. Besides, the standard model cannot account for the observed large fraction of dark mass of the universe. What is interesting is that in the universe the dark matter (DM) is about five times more abundant than the known baryonic matter, i.e. bright matter. We do not know why the ratio of dark to bright matter is of order unity.
- **Matter-Antimatter Asymmetry**: From our everyday experience we know that there is very little bright antimatter in the universe. The standard model fails to predict the observed excess of matter.

These arguments do not imply that the standard model is necessarily incorrect, but it must certainly be extended to answer any of the questions raised above. The truth is that we do not have an answer to the basic question: What lies beneath the standard model?

A number of possible generalizations of the standard model have been introduced based on one or more guiding principles or prejudices (see for technical reviews and the associated LHC phenomenology [13, 14]).

In the models we will consider here the electroweak symmetry breaks via a fermion bilinear condensate. The Higgs sector of the standard model becomes an effective description of a more fundamental fermionic theory. This is similar to the Ginzburg-Landau theory of superconductivity. If the force underlying the fermion condensate driving electroweak symmetry breaking is due to a strongly interacting gauge theory these models are termed technicolor.

Technicolor, in brief, is an additional non-abelian and strongly interacting gauge theory augmented with (techni)fermions transforming under a given representation of the gauge group. The Higgs Lagrangian is replaced by a suitable new fermion sector interacting strongly via a new gauge interaction (technicolor). Schematically:

$$L_{Higgs} \rightarrow -\frac{1}{4}F_{\mu\nu}F^{\mu\nu} + i\bar{Q}\gamma_\mu D^\mu Q + \cdots, \tag{1.14}$$

where, to be as general as possible, we have left unspecified the underlying nonabelian gauge group and the associated technifermion (Q) representation. The dots represent new sectors which may even be needed to avoid, for example, anomalies introduced by the technifermions. The intrinsic scale of the new theory is expected to be less or of the order of a few TeVs. The chiral-flavor symmetries of this theory, as for ordinary QCD, break spontaneously when the technifermion condensate $\bar{Q}Q$ forms. It is possible to choose the fermion charges in such a way that there is, at least, a weak left-handed doublet of technifermions and the associated right-handed one which is a weak singlet. The covariant derivative contains the new gauge field as well as the electroweak ones. The condensate spontaneously breaks the electroweak symmetry down to the electromagnetic and weak interactions. The Higgs is now interpreted as the lightest scalar field with the same quantum numbers of the fermion-antifermion composite field. The Lagrangian part responsible for the mass-generation of the ordinary fermions will also be modified since the Higgs particle is no longer an elementary object.

Models of electroweak symmetry breaking via new strongly interacting theories of technicolor type [15, 16] are a mature subject (for recent reviews see [17–19]). One of the main difficulties in constructing such extensions of the standard model is the very limited knowledge about generic strongly interacting theories. This has led theorists to consider specific models of technicolor which resemble ordinary quantum chromodynamics and for which the large body of experimental data at low energies can be directly exported to make predictions at high energies. Unfortunately the simplest version of this type of models are at odds with electroweak precision measurements. New strongly coupled theories with dynamics very different from the one featured by a scaled up version of QCD are needed [20].

We will review models of dynamical electroweak symmetry breaking making use of new type of four dimensional gauge theories [20–22] and their low energy effective description [23] useful for collider phenomenology. The phase structure of a large number of strongly interacting nonsupersymmetric theories, as function of number of underlying colors will be uncovered with traditional nonperturbative methods [24] as well as novel ones [25]. We will discuss possible applications to cosmology as well. These lectures should be integrated with earlier reviews [17–19, 26–31] on the various subjects treated here.

1.4 Superconductivity: The Condensed Matter Template

It is a fact that the standard model does not fail, when experimentally tested, to describe all of the known forces to a very high degree of experimental accuracy. This is true even if we include gravity. Why is it so successful? The standard model is a low energy effective theory valid up to a scale Λ above which new interactions, symmetries, extra dimensional worlds or any possible extension can emerge. At sufficiently low energies with respect to the cutoff scale Λ one expresses the existence of new physics via effective operators. The success of the standard model is due to the fact that most of the corrections to its physical observable depend only logarithmically on the cutoff scale Λ. Superrenormalizable operators are very sensitive to the cut off scale. In the standard model there exists only one operator with naive mass dimension two which acquires corrections quadratic in Λ. This is the squared mass operator of the Higgs boson. Since Λ is expected to be the highest possible scale, in four dimensions the Planck scale, it is hard to explain *naturally* why the mass of the Higgs is of the order of the electroweak scale. Due to the occurrence of quadratic corrections in the cutoff this is the standard model sector highly sensitive to the existence of new physics. In nature we have already observed Higgs-type mechanisms. Ordinary superconductivity and chiral symmetry breaking in QCD are two time-honored examples. In both cases the mechanism has an underlying dynamical origin with the Higgs-like particle being a composite object of fermionic fields. We will start from these two time-honored examples to motivate technicolor extensions of the standard model.

The breaking of the electroweak theory is a relativistic screening effect. It is useful to parallel it to ordinary superconductivity which is also a screening phenomenon albeit non-relativistic. The two phenomena happen at a temperature lower than a critical one. In the case of superconductivity one defines a density of superconductive electrons n_s and to it one associates a macroscopic wave function ψ such that its modulus squared

$$|\psi|^2 = n_C = \frac{n_s}{2}, \tag{1.15}$$

is the density of Cooper's pairs. That we are describing a nonrelativistic system is manifest in the fact that the macroscopic wave function squared, in natural units, has mass dimension three while the modulus squared of the Higgs wave function evaluated at the minimum is equal to $< |H|^2 >= v_{weak}^2/2$ and has mass dimension two, i.e. is a relativistic wave function. One can adjust the units by considering, instead of the wave functions, the Meissner-Mass of the photon in the superconductor which is

$$M^2 = q^2 n_s/(4m_e) , \tag{1.16}$$

with $q = -2e$ and $2m_e$ the charge and the mass of a Cooper pair which is consti-tuted by two electrons. In the electroweak theory the Meissner-Mass of the photon is compared with the relativistic mass of the W gauge boson

$$M_W^2 = g^2 v_{weak}^2/4 , \tag{1.17}$$

with g the weak coupling constant and v_{weak} the electroweak scale. In a superconduc-tor the relevant scale is given by the density of superconductive electrons typically of the order of $n_s \sim 4 \times 10^{28} \, m^{-3}$ yielding a screening length of the order of $\xi = 1/M \sim 10^{-6}$ cm. In the weak interaction case we measure directly the mass of the weak gauge boson which is of the order of $80 \, \text{GeV}$ yielding a weak screening length $\xi_W = 1/M_W \sim 10^{-15}$ cm.

For a superconductive system it is clear from the outset that the wave function ψ is not a fundamental degree of freedom, however for the Higgs we are not yet sure about its origin. The Ginzburg-Landau effective theory in terms of ψ and the photon degree of freedom describes the spontaneous breaking of the $U_Q(1)$ electric symmetry and it is the equivalent of the Higgs Lagrangian.

If the Higgs is due to a macroscopic relativistic screening phenomenon we expect it to be an effective description of a more fundamental system with possibly an underlying new strong gauge dynamics replacing the role of the phonons in the superconductive case. A dynamically generated Higgs system solves the problem of the quadratic divergences by replacing the cutoff Λ with the weak energy scale itself, i.e. the scale of compositeness. An underlying strongly coupled asymptotically free gauge theory, a la QCD, is an example.

1.5 From Color to Technicolor

Does the electroweak symmetry break in the complete absence of the standard model Higgs sector? The answer is *Yes*. QCD already breaks the electroweak symmetry spontaneously as we shall momentarily show. In fact, in QCD with two light quarks (up and down) the quantum global symmetry group is exactly $SU_L(2) \times SU_R(2)$ up to the baryon number $U_B(1)$. A matrix M_{QCD} of *composite* fields formally identical to the one introduced in the ordinary Higgs mechanism (e.g. (1.2)) can be constructed using the quarks bilinears which, in a suggestive form, reads

$$\sigma_{QCD} \sim \bar{q}q , \qquad \pi_{QCD} \sim i\bar{q}\tau\gamma_5 q, \tag{1.18}$$

building the QCD meson matrix M_{QCD}. This symmetry is known as the chiral symmetry of QCD which experiments have first indicated to break spontaneously to an unbroken subgroup $SU_V(2)$. In this case, when setting to zero the up and down quark masses, there are three Goldstone modes (π).

Turning on the electroweak interactions is done by simply recalling that the quarks up and down form a weak doublet. Therefore the $SU(2)_L$ symmetry is gauged by introducing the weak gauge bosons W^a with $a = 1, 2, 3$. The hypercharge generator is taken to be the third generator of $SU(2)_R$. The ordinary covariant derivative acting on the standard model Higgs (M) acts identically on the QCD matrix M_{QCD} and we have:

$$D_\mu M_{QCD} = \partial_\mu M_{QCD} - igW_\mu M_{QCD} + ig'M_{QCD} B_\mu, \quad \text{with} \quad W_\mu = W_\mu^a \frac{\tau^a}{2}, \ B_\mu = B_\mu \frac{\tau^3}{2}. \tag{1.19}$$

Recalling the fact that chiral symmetry is dynamically broken in QCD we have:

$$\langle \sigma_{QCD} \rangle^2 \equiv v_{QCD}^2 \quad \text{and} \quad \sigma_{QCD} = v_{QCD} + h_{QCD}, \tag{1.20}$$

where h_{QCD} is a QCD scalar meson. It is a bilinear, in the quarks, excitation around the vacuum expectation value of σ_{QCD} and remains heavy.[4]

In this way we have achieved that the global symmetry breaks dynamically to its diagonal subgroup:

$$SU_L(2) \times SU_R(2) \to SU_V(2). \tag{1.21}$$

To be more precise the $SU(2)_R$ symmetry is already broken explicitly by our choice of gauging only an $U_Y(1)$ subgroup of it and hence the actual symmetry breaking pattern is:

$$SU_L(2) \times U_Y(1) \to U_Q(1), \tag{1.22}$$

---

[4] The nature of this scalar state constitutes a phenomenological important puzzle in QCD. This is so since this state possesses the quantum numbers of the vacuum and therefore mixes with several other states of the theory made of higher quark and glue Fock states [32–36]. However there is a simple and clear limit in which some light can be shed. This is the 't Hooft mathematical large number of colors limit. In this limit all the mesons made by quark bilinears become stable and non-interacting. Therefore we expect this state to exist also when reducing the number of colors to three. One possibility is to identify it with $f_0(1300)$. Of course this state at any finite number of colors will never be a pure bilinear one. There is much confusion in the literature about this quark bilinear which is often erroneously identified with the $f_0(600)$ resonance. This state is not (mostly) a quark bilinear but its nature and physical properties are closer to the ones expected from a multi-quark state nature [32–34, 36]. $f_0(600)$ is, however, a crucial ingredient in the unitarization of pion-pion scattering at low energies and this fact is not directly related to the composition of the scalar meson but simply with its overall quantum numbers [32–34, 36] under space-time symmetries and the unbroken $SU_V(2)$ global symmetry.

with $U_Q(1)$ the electromagnetic abelian gauge symmetry. According to the Nambu-Goldstone's theorem three massless degrees of freedom appear, i.e. $\pi^{\pm}_{QCD}$ and π^0_{QCD}. In the unitary gauge these Goldstones become the longitudinal degree of freedom of the massive elecetroweak gauge-bosons. Substituting the vacuum value for σ_{QCD} in the Higgs Lagrangian the gauge-bosons quadratic terms read:

$$\frac{v^2_{QCD}}{8} \left[g^2 \left(W^1_\mu W^{\mu,1} + W^2_\mu W^{\mu,2} \right) + \left(g\, W^3_\mu - g' B_\mu \right)^2 \right]. \tag{1.23}$$

The Z_μ and the photon A_μ spectral relations are identical to the case of the standard model Higgs sector, and the masses are now $M^2_W = g^2 v^2_{QCD}/4$ which also due to the custodial symmetry (now naturally identified with the QCD isospin symmetry) satisfy the tree level relation $M^2_Z = M^2_W / \cos^2 \theta_W$.

What is missing to have a phenomenologically successful explanation of the observed spontaneous breaking of the electroweak gauge symmetry?

• The scale of the QCD condensate is too small to be able to account for the observed mass of the electroweak gauge bosons. In fact if QCD would be the only source contributing to the spontaneous breaking of the electroweak symmetry one would have

$$M_W = \frac{gF_\pi}{2} \sim 29\,\text{MeV}, \tag{1.24}$$

with $F_\pi = v_{QCD} \simeq 93\,\text{MeV}$ the pion decay constant. This contribution is very small with respect to the actual value of the M_W mass that one typically neglects it.
• We observed experimentally three light (pseudo) Goldstone bosons identified with the QCD pions.
• The quarks themselves have masses which means that another mechanism is in place for giving masses to the standard model fermions.

The first two issues can be resolved by postulating the existence of yet another strongly coupled theory with a dynamical scale taken to be the electroweak one ($v_{QCD} \rightarrow v_{weak}$), while the third one requires yet another sector. Weinberg [15] and Susskind [16] considered a new copy of QCD by simply rescaling the invariant mass of the theory, e.g. the new proton mass, and dubbed the model technicolor. However this model is at odds with experiments while the idea of dynamical breaking of the electroweak theory is very much alive. Whatever the new sector responsible for dynamical electroweak symmetry breaking is it will mix with the QCD one. In fact, the observed QCD physical pions are linear combination of the QCD pion eigenstates and the ones emerging from the new sector. The mixing is of the order of $v_{QCD}/v_{weak} \sim 10^{-3}$. Through this mixing the physical pion wave function, if measured with great accuracy, should be able to reveal the mechanism behind electroweak symmetry breaking using low energy data.

If the new gauge dynamics contains only fermionic matter than quantum corrections lead, at most, to logarithmic corrections to the bare parameters of the theory, and therefore the theory is said to be technically natural. A layman version of this statement is that: *small parameters stay small under renormalization*.

Therefore, according to the original idea of technicolor [15, 16] one augments the standard model with another gauge interaction similar to QCD but with a new dynamical scale of the order of the electroweak one. It is sufficient that the new gauge theory is asymptotically free and has global symmetry able to contain the standard model $SU_L(2) \times U_Y(1)$ symmetries. It is also required that the new global symmetries break dynamically in such a way that the embedded $SU_L(2) \times U_Y(1)$ breaks to the electromagnetic abelian charge $U_Q(1)$. The dynamically generated scale will then be fit to the electroweak one.

Note that, except in certain cases, dynamical behaviors are typically nonuniversal which means that different gauge groups and/or matter representations will, in general, possess very different dynamics.

The simplest example of technicolor theory is the scaled up version of QCD, i.e. an $SU(N_{TC})$ nonabelian gauge theory with two Dirac Fermions transforming according to the fundamental representation or the gauge group. We need at least two Dirac flavors to realize the $SU_L(2) \times SU_R(2)$ symmetry of the standard model discussed in the standard model Higgs section. One simply chooses the scale of the theory to be such that the new pion decaying constant is:

$$F_\pi^{TC} = v_{\text{weak}} \simeq 246 \, \text{GeV} . \tag{1.25}$$

The flavor symmetries, for any N_{TC} larger than 2 are $SU_L(2) \times SU_R(2) \times U_V(1)$ which spontaneously break to $SU_V(2) \times U_V(1)$. It is natural to embed the electroweak symmetries within the present technicolor model in such a way that the hypercharge corresponds to the third generator of $SU_R(2)$. This simple dynamical model correctly accounts for the electroweak symmetry breaking. The new technibaryon number $U_V(1)$ can break due to not yet specified new interactions. In order to get some indication on the dynamics and spectrum of this theory one can use the 't Hooft large N limit [37–39]. For example the intrinsic scale of the theory is related to the QCD one via:

$$\Lambda_{\text{TC}} \sim \sqrt{\frac{3}{N_{TC}}} \frac{F_\pi^{TC}}{F_\pi} \Lambda_{\text{QCD}} . \tag{1.26}$$

At this point it is straightforward to use the QCD phenomenology for describing the experimental signatures and dynamics of a composite Higgs.

1.6 Constraints from Electroweak Precision Data

The relevant corrections due to the presence of new physics trying to modify the electroweak breaking sector of the standard model appear in the vacuum polarizations of the electroweak gauge bosons. These can be parameterized in terms of the three quantities S, T, and U (the oblique parameters) [40–43], and confronted with the electroweak precision data. Recently, due to the increase precision of the measurements reported by LEP II, the list of interesting parameters to compute has been extended [44, 45]. We show below also the relation with the traditional one [40]. Defining with $Q^2 \equiv -q^2$ the Euclidean transferred momentum entering in a generic two point function vacuum polarization associated to the electroweak gauge bosons, and denoting derivatives with respect to $-Q^2$ with a prime we have [45]:

$$\hat{S} \equiv g^2 \, \Pi'_{W^3 B}(0), \tag{1.27}$$

$$\hat{T} \equiv \frac{g^2}{M_W^2} \left[\Pi_{W^3 W^3}(0) - \Pi_{W^+ W^-}(0) \right], \tag{1.28}$$

$$W \equiv \frac{g^2 M_W^2}{2} \left[\Pi''_{W^3 W^3}(0) \right], \tag{1.29}$$

$$Y \equiv \frac{g'^2 M_W^2}{2} \left[\Pi''_{BB}(0) \right], \tag{1.30}$$

$$\hat{U} \equiv -g^2 \left[\Pi'_{W^3 W^3}(0) - \Pi'_{W^+ W^-}(0) \right], \tag{1.31}$$

$$V \equiv \frac{g^2 M_W^2}{2} \left[\Pi''_{W^3 W^3}(0) - \Pi''_{W^+ W^-}(0) \right], \tag{1.32}$$

$$X \equiv \frac{g g' M_W^2}{2} \, \Pi''_{W^3 B}(0). \tag{1.33}$$

Here $\Pi_V(Q^2)$ with $V = \{W^3 B, W^3 W^3, W^+ W^-, BB\}$ represents the self-energy of the vector bosons. The electroweak couplings are the ones associated to the physical electroweak gauge bosons:

$$\frac{1}{g^2} \equiv \Pi'_{W^+ W^-}(0), \qquad \frac{1}{g'^2} \equiv \Pi'_{BB}(0), \tag{1.34}$$

while G_F is

$$\frac{1}{\sqrt{2} G_F} = -4 \Pi_{W^+ W^-}(0), \tag{1.35}$$

as in [46]. $\hat{S}$ and $\hat{T}$ lend their name from the well known Peskin-Takeuchi parameters S and T which are related to the new ones via [45, 46]:

$$\frac{\alpha S}{4 s_W^2} = \hat{S} - Y - W, \qquad \alpha T = \hat{T} - \frac{s_W^2}{1 - s_W^2} Y. \tag{1.36}$$

Here α is the electromagnetic structure constant and $s_W = \sin \theta_W$ is the weak mixing angle. Therefore in the case where $W = Y = 0$ we have the simple relation

$$\hat{S} = \frac{\alpha S}{4 s_W^2}, \qquad \hat{T} = \alpha T. \tag{1.37}$$

Fig. 1.7 *T* versus *S* for
$SU(3)$ technicolor with one
technifermion doublet (the
black point) versus precision
data for a one TeV composite
Higgs mass (the *shaded area*)

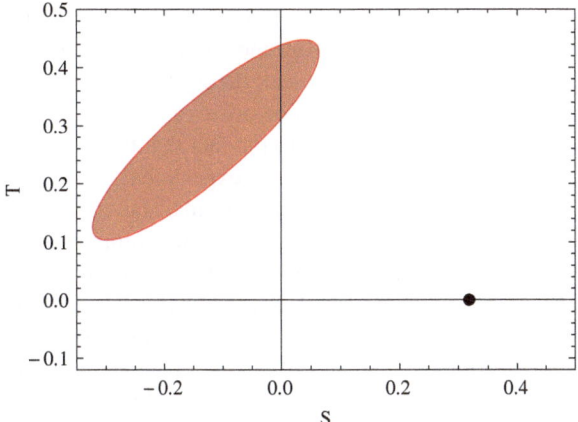

The result of the fit is shown in Fig. 1.5. If the value of the Higgs mass increases the central value of the *S* parameter moves to the left towards negative values.

In technicolor it is easy to have a vanishing *T* parameter while typically *S* is positive. Besides, the composite Higgs is typically heavy with respect to the Fermi scale, at least for technifermions in the fundamental representation of the gauge group and for a small number of techniflavors. The oldest technicolor models featuring QCD dynamics with three technicolors and a doublet of electroweak gauged technifla-vors deviate a few sigma from the current precision tests as summarized in Fig. 1.7. Clearly it is desirable to reduce the tension between the precision data and a possible dynamical mechanism underlying the electroweak symmetry breaking. It is possible to imagine different ways to achieve this goal and some of the earlier attempts have been summarized in [47].

The computation of the *S* parameter in technicolor theories requires the knowledge of nonperturbative dynamics making difficult the precise knowledge of the contribu-tion to *S*. For example, it is not clear what is the exact value of the composite Higgs mass relative to the Fermi scale and, to be on the safe side, one typically takes it to be quite large, of the order at least of the TeV. However in certain models it may be substantially lighter due to the intrinsic dynamics. We will discuss the electroweak parameters later in this chapter.

It is, however, instructive to provide a simple estimate of the contribution to *S* which allows to guide model builders. Consider a one-loop exchange of N_D doublets of techniquarks transforming according to the representation R_{TC} of the underlying technicolor gauge theory and with dynamically generated mass $\Sigma(0)$ assumed to be larger than the weak intermediate gauge bosons masses. Indicating with $d(R_{TC})$ the dimension of the techniquark representation, and to leading order in $M_W/\Sigma(0)$ one finds:

$$S_{\text{naive}} = N_D \frac{d(R_{TC})}{6\pi} . \tag{1.38}$$

This naive value provides, in general, only a rough estimate of the exact value of S. However, it is clear from the formula above that, the more technicolor matter is gauged under the electroweak theory the larger is the S parameter and that the final S parameter is expected to be positive.

Attention must be paid to the fact that the specific model-estimate of the whole S parameter, to compare with the experimental value, receives contributions also from other sectors. Such a contribution can be taken sufficiently large and negative to compensate for the positive value from the composite Higgs dynamics. To be concrete: Consider an extension of the standard model in which the Higgs is composite but we also have new heavy (with a mass of the order of the electroweak) fourth family of Dirac leptons. In this case a sufficiently large splitting of the new lepton masses can strongly reduce and even offset the positive value of S. We will discuss this case in detail when presenting the Minimal Walking technicolor (MWT) model. The contribution of the new sector (S_{NS}) above, and also in many other cases, is perturbatively under control and the total S can be written as:

$$S = S_{TC} + S_{NS} \,. \tag{1.39}$$

The parameter T will be, in general, modified and one has to make sure that the corrections do not spoil the agreement with this parameter. From the discussion above it is clear that technicolor models can be constrained, via precision measurements, only model by model and the effects of possible new sectors must be properly included. We presented the constraints coming from S using the underlying gauge theory information. However, in practice, these constraints apply directly to the physical spectrum. To be concrete we will present in Sect. 3.3.7 a model of walking technicolor passing the precision tests.

1.7 Standard Model Fermion Masses

Since in a purely technicolor model the Higgs is a composite particle the Yukawa terms, when written in terms of the underlying technicolor fields, amount to four-fermion operators. The latter can be naturally interpreted as a low energy operator induced by a new strongly coupled gauge interaction emerging at energies higher than the electroweak theory. These type of theories have been termed Extended technicolor (ETC) interactions [48, 49].

In the literature various extensions have been considered and we will mention them later in the text. Here we will describe the simplest ETC model in which the ETC interactions connect the chiral symmetries of the techniquarks to those of the standard model fermions (see left panel of Fig. 1.8).

When technicolor chiral symmetry breaking occurs it leads to the diagram in the right panel of Fig. 1.8. Let's start with the case in which the ETC dynamics is represented by a $SU(N_{ETC})$ gauge group with:

$$N_{ETC} = N_{TC} + N_g \,, \tag{1.40}$$

Fig. 1.8 *Left panel* ETC gauge boson interaction involving techniquarks and standard model fermions. *Right panel* diagram contribution to the mass to the standard model fermions

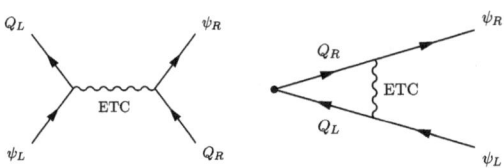

and N_g is the number of standard model generations. In order to give masses to all of the standard model fermions, in this scheme, one needs a condensate for each standard model fermion. This can be achieved by using as technifermion matter a complete generation of quarks and leptons (including a neutrino right) but now gauged with respect to the technicolor interactions.

The ETC gauge group is assumed to spontaneously break N_g times down to $SU(N_{TC})$ permitting three different mass scales, one for each standard model family. This type of technicolor with associated ETC is termed the *one family model* [50]. The heavy masses are provided by the breaking at low energy and the light masses are provided by breaking at higher energy scales. This model does not, per se, explain how the gauge group is broken several times, neither is the breaking of weak isospin symmetry accounted for. For example we cannot explain why the neutrino have masses much smaller than the associated electrons. See, however, [51] for progress on these issues. Schematically one has $SU(N_{TC} + 3)$ which breaks to $SU(N_{TC} + 2)$ at the scale Λ_1 providing the first generation of fermions with a typical mass $m_1 \sim 4\pi(F_\pi^{TC})^3/\Lambda_1^2$. At this point the gauge group breaks to $SU(N_{TC} + 1)$ with dynamical scale Λ_2 leading to a second generation mass of the order of $m_2 \sim 4\pi(F_\pi^{TC})^3/\Lambda_2^2$. Finally the last breaking $SU(N_{TC})$ at scale Λ_3 leading to the last generation mass $m_3 \sim 4\pi(F_\pi^{TC})^3/\Lambda_3^2$.

Without specifying an ETC one can write down the most general type of four-fermion operators involving technicolor particles Q and ordinary fermionic fields ψ. Following the notations of Hill and Simmons [17] we write:

$$\alpha_{ab}\frac{\bar{Q}\gamma_\mu T^a Q\bar{\psi}\gamma^\mu T^b\psi}{\Lambda_{ETC}^2} + \beta_{ab}\frac{\bar{Q}\gamma_\mu T^a Q\bar{Q}\gamma^\mu T^b Q}{\Lambda_{ETC}^2} + \gamma_{ab}\frac{\bar{\psi}\gamma_\mu T^a\psi\bar{\psi}\gamma^\mu T^b\psi}{\Lambda_{ETC}^2}, \quad (1.41)$$

where the Ts are unspecified ETC generators. After performing a Fierz rearrangement one has:

$$\alpha_{ab}\frac{\bar{Q}T^a Q\bar{\psi}T^b\psi}{\Lambda_{ETC}^2} + \beta_{ab}\frac{\bar{Q}T^a Q\bar{Q}T^b Q}{\Lambda_{ETC}^2} + \gamma_{ab}\frac{\bar{\psi}T^a\psi\bar{\psi}T^b\psi}{\Lambda_{ETC}^2} + \cdots, \quad (1.42)$$

The coefficients parameterize the ignorance on the specific ETC physics. To be more specific, the α-terms, after the technicolor particles have condensed, lead to mass terms for the standard model fermions

$$m_q \approx \frac{g_{ETC}^2}{M_{ETC}^2}\langle\bar{Q}Q\rangle_{ETC}, \quad (1.43)$$

Fig. 1.9 Leading contribution
to the mass of the technicolor
pseudo Goldstone bosons via
an exchange of an ETC gauge
boson

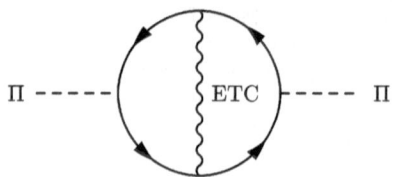

where m_q is the mass of e.g. a standard model quark, g_{ETC} is the ETC gauge coupling
constant evaluated at the ETC scale, M_{ETC} is the mass of an ETC gauge boson and
$\langle \bar{Q}Q \rangle_{ETC}$ is the technicolor condensate where the operator is evaluated at the ETC
scale. Note that we have not explicitly considered the different scales for the different
generations of ordinary fermions but this should be taken into account for any realistic
model.

The β-terms of Eq. (1.42) provide masses for pseudo Goldstone bosons and also
provide masses for techniaxions [17], see Fig. 1.9. The last class of terms, namely
the γ-terms of Eq. (1.42) induce FCNCs. For example it may generate the following
terms:

$$\frac{1}{\Lambda_{ETC}^2}(\bar{s}\gamma^5 d)(\bar{s}\gamma^5 d) + \frac{1}{\Lambda_{ETC}^2}(\bar{\mu}\gamma^5 e)(\bar{e}\gamma^5 e) + \cdots , \qquad (1.44)$$

where s, d, μ, e denote the strange and down quark, the muon and the electron,
respectively. The first term is a $\Delta S = 2$ flavor-changing neutral current interaction
affecting the $K_L - K_S$ mass difference which is measured accurately. The experimental
bounds on these type of operators, together with the very *naive* assumption that ETC
will generate γ-terms with coefficients of order one, leads to a constraint on the ETC
scale to be of the order of or larger than 10^3 TeV [48]. This should be the lightest ETC
scale which in turn puts an upper limit on how large the ordinary fermionic masses
can be. The naive estimate is that one can account up to around 100 MeV mass for
a QCD-like technicolor theory, implying that the top quark mass value cannot be
achieved.

The second term of Eq. (1.44) induces flavor changing processes in the leptonic
sector such as $\mu \to e\bar{e}e, e\gamma$ which are not observed. It is clear that, both for the
precision measurements and the fermion masses, a better theory of the flavor is
needed. For the ETC dynamics interesting developments recently appeared in the
literature [52–55]. We note that nonperturbative chiral gauge theories dynamics is
expected to play a relevant role in models of ETC since it allows, at least in principle,
the self breaking of the gauge symmetry. Recent progress on the phase diagrams of
these theories has appeared in [56].

In Fig. 1.10 we show the ordering of the relevant scales involved in the generation
of the ordinary fermion masses via ETC dynamics, and the generation of the fermion
masses (for a single generation and focussing on the top quark) assuming QCD-like
dynamics for technicolor.

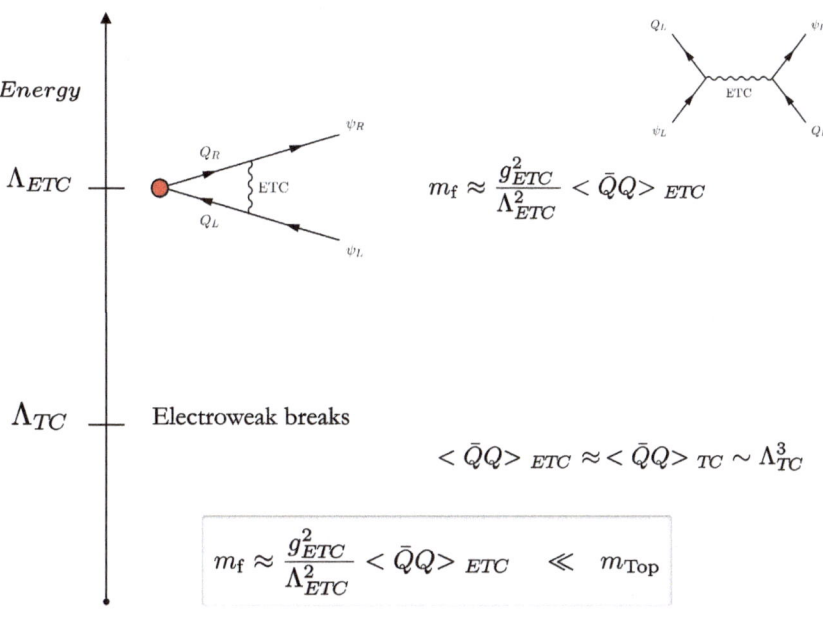

Fig. 1.10 Cartoon of the expected ETC dynamics starting at high energies with a more fundamental gauge interaction and the generation of the fermion masses assuming QCD-like dynamics

1.8 Walking

To better understand in which direction one should go to modify the QCD dynamics, we analyze the technicolor condensate. The value of the TC condensate used when giving mass to the ordinary fermions should be evaluated not at the technicolor scale but at the ETC one. Via the renormalization group one can relate the condensate at the two scales via:

$$\langle \bar{Q}Q \rangle_{\text{ETC}} = \exp \left(\int_{\Lambda_{\text{TC}}}^{\Lambda_{\text{ETC}}} d(\ln \mu) \gamma(\alpha(\mu)) \right) \langle \bar{Q}Q \rangle_{\text{TC}}, \qquad (1.45)$$

where γ is the anomalous dimension of the techniquark mass-operator. The boundaries of the integral are at the ETC scale and the technicolor one. For technicolor theories with a running of the coupling constant similar to the one in QCD, i.e.

$$\alpha(\mu) \propto \frac{1}{\ln \mu}, \quad \text{for } \mu > \Lambda_{\text{TC}}, \qquad (1.46)$$

this implies that the anomalous dimension of the techniquark masses $\gamma \propto \alpha(\mu)$. When computing the integral one gets

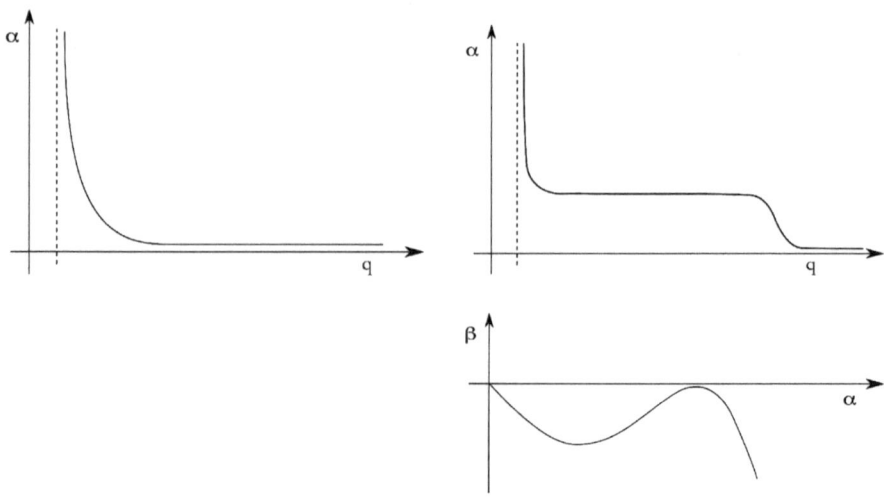

Fig. 1.11 *Top left panel* QCD-like behavior of the coupling constant as function of the momentum (Running). *Top right panel* walking-like behavior of the coupling constant as function of the momentum (Walking). *Bottom right panel* cartoon of the beta function associated to a generic walking theory

$$\langle \bar{Q}Q \rangle_{\text{ETC}} \sim \ln \left(\frac{\Lambda_{\text{ETC}}}{\Lambda_{\text{TC}}} \right)^{\gamma} \langle \bar{Q}Q \rangle_{\text{TC}}, \tag{1.47}$$

which is a logarithmic enhancement of the operator. We can hence neglect this correction and use directly the value of the condensate at the technicolor scale when estimating the generated fermionic mass:

$$m_q \approx \frac{g_{\text{ETC}}^2}{M_{\text{ETC}}^2} \Lambda_{\text{TC}}^3, \qquad \langle \bar{Q}Q \rangle_{\text{TC}} \sim \Lambda_{\text{TC}}^3. \tag{1.48}$$

The tension between having to reduce the FCNCs and at the same time provide a sufficiently large mass for the heavy fermions in the standard model as well as the pseudo-Goldstones can be reduced if the dynamics of the underlying technicolor theory is different from the one of QCD. The computation of the technicolor condensate at different scales shows that if the dynamics is such that the technicolor coupling does not *run* to the UV fixed point but rather slowly reduces to zero one achieves a net enhancement of the condensate itself with respect to the value estimated earlier. This can be achieved if the theory has a near conformal fixed point. This kind of dynamics has been denoted as of *walking* type. In Fig. 1.11 the comparison between a running and walking behavior of the coupling is qualitatively represented.

In the walking regime:

$$\langle \bar{Q}Q \rangle_{\text{ETC}} \sim \left(\frac{\Lambda_{\text{ETC}}}{\Lambda_{\text{TC}}} \right)^{\gamma(\alpha^*)} \langle \bar{Q}Q \rangle_{\text{TC}} \,, \tag{1.49}$$

which is a much larger contribution than in QCD dynamics [57–60]. Here γ is evaluated at the would be fixed point value α^*. Walking can help resolving the problem of FCNCs in technicolor models since with a large enhancement of the $\langle \bar{Q}Q \rangle$ condensate the four-Fermi operators involving standard model fermions and technifermions and the ones involving technifermions are enhanced by a factor of $\Lambda_{\text{ETC}}/\Lambda_{\text{TC}}$ to the γ power while the one involving only standard model fermions is not enhanced.

We note that *walking* is not a fundamental property for a successful model of the origin of mass of the elementary fermions featuring technicolor. In fact several alternative ideas already exist in the literature (see [61–64] for recent models while for earlier models we refer to [65–75]). However, a near conformal theory would still be useful to reduce the contributions to the precision data and, possibly, provide a light composite Higgs of much interest to LHC physics [21].

1.9 Ideal Walking

There are several issues associated with the original idea of walking:

- Since the number of flavors cannot be changed continuously it is not possible to get arbitrarily close to the lower end of the conformal window. This applies to the technicolor theory *in isolation* i.e. before coupling it to the standard model and without taking into account the ETC interactions.
- It is hard to achieve large anomalous dimensions of the fermion mass operator even near the lower end of the conformal window for ordinary gauge theories.
- It is not always possible to neglect the interplay of the four fermion interactions on the technicolor dynamics.
- There exist the logical possibility that, as function of the number of flavors, the theory jumps out of the conformal window rather than walk out of it. Jumping dynamics [76] will be discussed in the next chapter.

In [77] it has been argued that it is possible to address simultaneously the problems above, expect for the possibility of jumping, by taking into account the effects of the four-fermion interactions on the phase diagram of strongly interacting theories for any matter representation as function of the number of colors and flavors. A positive effect is that the anomalous dimension of the mass increases beyond the unity value at the lower boundary of the new conformal window and can get sufficiently large to yield the correct mass for the top quark. It has also been shown that the conformal window, for any representation, shrinks with respect to the case in which the four-fermion interactions are neglected. This analysis derives from the study of the gauged Nambu-Jona-Lasinio phase diagram [78].

It has been made the further discovery that when the extended technicolor sector, responsible for giving masses to the standard model fermions, is sufficiently strongly coupled, the technicolor theory, in isolation, must feature an infrared fixed point in order for the full model to be phenomenologically viable and correctly break the electroweak symmetry [77].

1.10 Walking Spectrum

Any strongly coupled dynamics, even of walking type, will generate a spectrum of resonances whose natural splitting in mass is of the order of the intrinsic scale of the theory which in this case is the Fermi scale. In order to extract predictions for the composite vector spectrum and couplings in presence of a strongly interacting sector and an asymptotically free gauge theory, we make use of the time-honored Weinberg sum rules [79] but we will also use the results found in [80] allowing us to treat walking and running theories in a unified way.

1.10.1 Weinberg Sum Rules

The Weinberg sum rules are linked to the two point vector–vector minus axial–axial vacuum polarization which is known to be sensitive to chiral symmetry breaking. We define

$$
i\Pi_{\mu\nu}^{a,b}(q) \equiv \int d^4x \, e^{-iqx} \left[< J_{\mu,V}^a(x) J_{\nu,V}^b(0) > \; - \; < J_{\mu,A}^a(x) J_{\nu,A}^b(0) > \right], \quad (1.50)
$$

within the underlying strongly coupled gauge theory, where

$$
\Pi_{\mu\nu}^{a,b}(q) = \left(q_\mu q_\nu - g_{\mu\nu} q^2 \right), \delta^{ab} \Pi(q^2). \quad (1.51)
$$

Here $a, b = 1, \ldots, n_f^2 - 1$, label the flavor currents and the $SU(N_f)$ generators are normalized according to $\mathrm{Tr}\left[T^a T^b \right] = (1/2)^{ab}$. The function $\Pi(q^2)$ obeys the unsubtracted dispersion relation

$$
\frac{1}{\pi} \int_0^\infty ds \, \frac{\mathrm{Im}\Pi(s)}{s + Q^2} = \Pi(Q^2), \quad (1.52)
$$

where $Q^2 = -q^2 > 0$, and the constraint $-Q^2 \Pi(Q^2) > 0$ holds for $0 < Q^2 < \infty$ [81]. The discussion above is for the standard chiral symmetry breaking pattern $SU(N_f) \times SU(N_f) \to SU(N_f)$ but it is generalizable to any breaking pattern.

Since we are taking the underlying theory to be asymptotically free, the behavior of $\Pi(Q^2)$ at asymptotically high momenta is the same as in ordinary QCD, i.e. it scales like Q^{-6} [82]. Expanding the left hand side of the dispersion relation thus leads to the two conventional spectral function sum rules

$$\frac{1}{\pi} \int_0^\infty ds \mathrm{Im} \Pi(s) = 0 \quad \text{and} \quad \frac{1}{\pi} \int_0^\infty ds\, s\, \mathrm{Im} \Pi(s) = 0. \tag{1.53}$$

Walking dynamics affects only the second sum rule [80] which is more sensitive to large but not asymptotically large momenta due to fact that the associated integrand contains an extra power of s.

We now saturate the absorptive part of the vacuum polarization. We follow reference [80] and hence divide the energy range of integration in three parts. The light resonance part. In this regime, the integral is saturated by the Nambu-Goldstone pseudoscalar along with massive vector and axial–vector states. If we assume, for example, that there is only a single, zero-width vector multiplet and a single, zero-width axial vector multiplet, then

$$\mathrm{Im} \Pi(s) = \pi F_V^2 \delta\left(s - M_V^2\right) - \pi F_A^2 \delta\left(s - M_A^2\right) - \pi F_\pi^2 \delta(s). \tag{1.54}$$

The zero-width approximation is valid to leading order in the large N expansion for fermions in the fundamental representation of the gauge group and it is even narrower for fermions in higher dimensional representations. Since we are working near a conformal fixed point the large N argument for the width is not directly applicable. We will nevertheless use this simple model for the spectrum to illustrate the effects of a near critical IR fixed point.

The first Weinberg sum rule implies:

$$F_V^2 - F_A^2 = F_\pi^2, \tag{1.55}$$

where F_V^2 and F_A^2 are the vector and axial mesons decay constants. This sum rule holds for walking and running dynamics. A more general representation of the resonance spectrum would, in principle, replace the left hand side of this relation with a sum over vector and axial states. However the heavier resonances should not be included since in the approach of [80] the walking dynamics in the intermediate energy range is already approximated by the exchange of underlying fermions. The walking is encapsulated in the dynamical mass dependence on the momentum dictated by the gauge theory. The introduction of heavier resonances is, in practice, double counting. Note that the approach is in excellent agreement with the Weinberg approximation for QCD, since in this case, the approximation automatically returns the known results.

The second sum rule receives important contributions from throughout the near conformal region and can be expressed in the form of:

$$F_V^2 M_V^2 - F_A^2 M_A^2 = a \frac{8\pi^2}{d(R)} F_\pi^4, \tag{1.56}$$

where a is expected to be positive and $O(1)$ and $d(R)$ is the dimension of the representation of the underlying fermions. We have generalized the result of reference [80] to the case in which the fermions belong to a generic representation of the gauge group. In the case of running dynamics the right-hand side of the previous equation vanishes.

We stress that a is a non-universal quantity depending on the details of the underlying gauge theory. A reasonable measure of how large a can be is given by a function of the amount of walking which is the ratio of the scale above which the underlying coupling constant start running divided by the scale below which chiral symmetry breaks. The fact that a is positive and of order one in walking dynamics is supported, indirectly, also via the results of Kurachi and Shrock [83]. At the onset of conformal dynamics the axial and the vector will be degenerate, i.e. $M_A = M_V = M$, using the first sum rule one finds via the second sum rule $a = d(R)M^2/(8\pi^2 F_\pi^2)$ leading to a numerical value of about 4–5 from the approximate results in [83]. We will however use only the constraints coming from the generalized Weinberg sum rules expecting them to be less model dependent. The S parameter is related to the absorptive part of the vector–vector minus axial–axial vacuum polarization as follows:

$$S = 4 \int_0^\infty \frac{ds}{s} \mathrm{Im}\bar{\Pi}(s) = 4\pi \left[\frac{F_V^2}{M_V^2} - \frac{F_A^2}{M_A^2} \right], \qquad (1.57)$$

where $\mathrm{Im}\bar{\Pi}$ is obtained from $\mathrm{Im}\Pi$ by subtracting the Goldstone boson contribution.

Other attempts to estimate the S parameter for walking technicolor theories have been made in the past [84] showing reduction of the S parameter. S has also been evaluated using computations inspired by the original AdS/CFT correspondence [85] in [86–91]. Recent attempts to use AdS/CFT inspired methods can be found in [92–96].

Kurachi, Shrock and Yamawaki [97] have further confirmed the results presented in [80] with their computations tailored for describing four dimensional gauge theories near the conformal window. The present approach [80] is more physical since it is based on the nature of the spectrum of states associated directly to the underlying gauge theory.

Note that we will be assuming a rather conservative approach in which the S parameter, although reduced with respect to the case of a running theory, is bounded by the naive S parameter [98–100]. After all, other sectors of the theory such as new leptons further reduce or completely offset a positive value of S due solely to the technicolor theory.

References

1. H. Flacher, M. Goebel, J. Haller, A. Hocker, K. Monig, J. Stelzer, Eur. Phys. J. C **60**, 543 (2009) [Erratum-ibid. C **71**, 1718 (2011)] [arXiv:0811.0009 [hep-ph]
2. G. Aad et al. [ATLAS Collaboration], Observation of a new particle in the search for the standard model Higgs boson with the ATLAS detector at the LHC, arXiv:1207.7214 [hep-ex]

3. S. Chatrchyan et al. [CMS Collaboration], Observation of a new boson at a mass of 125 GeV with the CMS experiment at the LHC, [arXiv:1207.7235 [hep-ex]]
4. C. a. D. C. a. t. T. N. P. a. H. W. Group [Tevatron New Physics Higgs Working Group and CDF and D0 Collaborations], Updated Combination of CDF and D0 Searches for Standard Model Higgs Boson Production with up to 10.0 fb^{-1} of Data, arXiv:1207.0449 [hep-ex]
5. ATLAS Collaboration, (2012), 1202.1408
6. C.M.S. Collaboration, S. Chatrchyan et al. (2012), 1202.1488
7. *An update to the combined search for the standard model higgs boson with the atlas detector at the lhc using up to 4.9 ?1 of pp collision data at +s = 7 tev*, Technical Report ATLAS-CONF-2012-019, CERN, Geneva, March 2012
8. *Combined results of searches for a higgs boson in the context of the standard model and beyond-standard models*, Technical Report CMS-PAS-HIG-12-008, CERN, Geneva, 2012
9. S. Kortner, Sm scalar boson search with the atlas detector, Moriond 2012
10. M. Pieri, Searches for the sm scalar boson at cms
11. R. Barate et al., LEP Working Group for Higgs boson searches, ALEPH Collaboration, DELPHI Collaboration, L3 Collaboration, OPAL Collaboration. Phys. Lett. **B565**, 61–75 (2003), [hep-ex/0306033]
12. F. Sannino, Mod. Phys. Lett. A **26**, 1763 (2011) [arXiv:1102.5100 [hep-ph]]
13. A. De Roeck et al., arXiv:0909.3240 [Unknown]
14. E. Accomando et al., arXiv:hep-ph/0608079
15. S. Weinberg, Phys. Rev. **D19**, 1277 (1979)
16. L. Susskind, Phys. Rev. **D20**, 2619 (1979)
17. C.T. Hill, E.H. Simmons, Phys. Rept. **381**, 235 (2003) [Erratum-ibid. **390**, 553 (2004)] [arXiv:hep-ph/0203079]
18. F. Sannino, arXiv:0804.0182 [hep-ph]
19. K. Lane, arXiv:hep-ph/0202255
20. F. Sannino, K. Tuominen, Phys. Rev. **D71**, 051901 (2005) [arXiv:hep-ph/0405209]
21. D.D. Dietrich, F. Sannino, K. Tuominen, Phys. Rev. **D72**, 055001 (2005) [arXiv:hep-ph/0505059]
22. D.D. Dietrich, F. Sannino, K. Tuominen, Phys. Rev. **D73**, 037701 (2006) [arXiv:hep-ph/0510217]
23. R. Foadi, M. T. Frandsen, T. A. Ryttov, F. Sannino, Phys. Rev. D 76, 055005 (2007) [arXiv:0706.1696 [hep-ph]]
24. D.D. Dietrich, F. Sannino, Phys. Rev. **D75**, 085018 (2007) [arXiv:hep-ph/0611341]
25. T.A. Ryttov, F. Sannino, Phys. Rev. **D78**, 065001 (2008) [arXiv:0711.3745 [hep-th]]
26. R. Shrock, arXiv:hep-ph/0703050
27. S. Sarkar, Rept. Prog. Phys. **59**, 1493 (1996) [arXiv:hep-ph/9602260]
28. M.S. Chanowitz, Ann. Rev. Nucl. Part. Sci. **38**, 323 (1988)
29. E. Farhi, L. Susskind, Phys. Rept. **74**, 277 (1981)
30. R.K. Kaul, Rev. Mod. Phys. **55**, 449 (1983)
31. R.S. Chivukula, arXiv:hep-ph/0011264
32. F. Sannino, J. Schechter, Phys. Rev. **D52**, 96 (1995) [arXiv:hep-ph/9501417]
33. M. Harada, F. Sannino, J. Schechter, Phys. Rev. **D54**, 1991 (1996) [arXiv:hep-ph/9511335]
34. M. Harada, F. Sannino, J. Schechter, Phys. Rev. **D69**, 034005 (2004) [arXiv:hep-ph/0309206]
35. J.R. Pelaez, Phys. Rev. Lett. **92**, 102001 (2004) [arXiv:hep-ph/0309292]
36. F. Sannino, J. Schechter, Phys. Rev. **D76**, 014014 (2007) [arXiv:0704.0602 [hep-ph]]
37. G. 't Hooft, Nucl. Phys. B **72**, 461 (1974)
38. E. Witten, Nucl. Phys. B **160**, 57 (1979)
39. G. 't Hooft, C. Itzykson, A. Jaffe, H. Lehmann, P.K. Mitter, I.M. Singer, R. Stora (eds.), *Recent Development in Gauge Theories*. Nato Advanced Study Institutes Series: Series B, Physics, vol. 59 (Plenum, New York, 1980), p. 438
40. M.E. Peskin, T. Takeuchi, Phys. Rev. Lett. **65**, 964 (1990)
41. M.E. Peskin, T. Takeuchi, Phys. Rev. **D46**, 381 (1992)
42. D.C. Kennedy, P. Langacker, Phys. Rev. Lett. **65**, 2967 (1990) [Erratum-ibid. 66, 395 (1991)]

43. G. Altarelli, R. Barbieri, Phys. Lett. B **253**, 161 (1991)
44. I. Maksymyk, C.P. Burgess, D. London, Phys. Rev. **D50**, 529 (1994) [arXiv:hep-ph/9306267]
45. R. Barbieri, A. Pomarol, R. Rattazzi, A. Strumia, Nucl. Phys. B **703**, 127 (2004) [arXiv:hep-ph/0405040]
46. R.S. Chivukula, E.H. Simmons, H.J. He, M. Kurachi, M. Tanabashi, Phys. Lett. B **603**, 210 (2004) [arXiv:hep-ph/0408262]
47. M.E. Peskin, J.D. Wells, Phys. Rev. **D64**, 093003 (2001) [arXiv:hep-ph/0101342]
48. E. Eichten, K.D. Lane, Phys. Lett. B **90**, 125 (1980)
49. S. Dimopoulos, L. Susskind, Nucl. Phys. B **155**, 237 (1979)
50. E. Farhi, L. Susskind, Phys. Rev. **D20**, 3404 (1979)
51. T. Appelquist, N.D. Christensen, M. Piai, R. Shrock, Phys. Rev. **D70**, 093010 (2004) [arXiv:hep-ph/0409035]
52. T.A. Ryttov, R. Shrock, Phys. Rev. **D81**, 115013 (2010) [arXiv:1004.2075 [hep-ph]]
53. T.A. Ryttov, R. Shrock, Eur. Phys. J. C **71**, 1523 (2011) [arXiv:1005.3844 [hep-ph]]
54. T.A. Ryttov, R. Shrock, Phys. Rev. **D82**, 055012 (2010) [arXiv:1006.5477 [hep-ph]]
55. N. Chen, T.A. Ryttov, R. Shrock, Phys. Rev. **D82**, 116006 (2010) [arXiv:1010.3736 [hep-ph]]
56. F. Sannino, Acta Phys. Polon. B **40**, 3533 (2009) [arXiv:0911.0931 [hep-ph]]
57. K. Yamawaki, M. Bando, K. Matumoto, Phys. Rev. Lett. **56**, 1335 (1986)
58. B. Holdom, Phys. Lett. B **150**, 301 (1985)
59. B. Holdom, Phys. Rev. **D24**, 1441 (1981)
60. T.W. Appelquist, D. Karabali, L.C.R. Wijewardhana, Phys. Rev. Lett. **57**, 957 (1986)
61. M. Antola, M. Heikinheimo, F. Sannino, K. Tuominen, arXiv:0910.3681 [hep-ph]
62. M. Antola, S. Di Chiara, F. Sannino, K. Tuominen, Eur. Phys. J. C **71**, 1784 (2011) [arXiv:1001.2040 [hep-ph]]
63. M. Antola, S. Di Chiara, F. Sannino, K. Tuominen, Nucl. Phys. B **856**, 647 (2012) [arXiv:1009.1624 [hep-ph]]
64. M. Antola, S. Di Chiara, F. Sannino, K. Tuominen, arXiv:1111.1009 [hep-ph]
65. E.H. Simmons, Nucl. Phys. B **312**, 253 (1989)
66. M. Dine, A. Kagan, S. Samuel, Phys. Lett. B **243**, 250 (1990)
67. S. Samuel, Nucl. Phys. B **347**, 625 (1990)
68. A. Kagan, S. Samuel, Phys. Lett. B **270**, 37 (1991)
69. A. Kagan, S. Samuel, Phys. Lett. B **284**, 289 (1992)
70. A. Kagan, S. Samuel, Int. J. Mod. Phys. A **7**, 1123 (1992)
71. C.D. Carone, E.H. Simmons, Nucl. Phys. B **397**, 591 (1993) [arXiv:hep-ph/9207273]
72. C.D. Carone, E.H. Simmons, Y. Su, Phys. Lett. B **344**, 287 (1995) [arXiv:hep-ph/9410242]
73. V. Hemmige, E.H. Simmons, Phys. Lett. B **518**, 72 (2001) [arXiv:hep-ph/0107117]
74. C.D. Carone, J. Erlich, J.A. Tan, Phys. Rev. **D75**, 075005 (2007) [arXiv:hep-ph/0612242]
75. A.R. Zerwekh, arXiv:0907.4690 [hep-ph]
76. F. Sannino, arXiv:1205.4246 [hep-ph]
77. H.S. Fukano, F. Sannino, Phys. Rev. **D82**, 035021 (2010) [arXiv:1005.3340 [hep-ph]]
78. K. Kondo, M. Tanabashi, K. Yamawaki, Mod. Phys. Lett. A **8**, 2859 (1993)
79. S. Weinberg, Phys. Rev. Lett. **18**, 507 (1967)
80. T. Appelquist, F. Sannino, Phys. Rev. **D59**, 067702 (1999) [arXiv:hep-ph/9806409]
81. E. Witten, Phys. Rev. Lett. **51**, 2351 (1983)
82. C.W. Bernard, A. Duncan, J. LoSecco, S. Weinberg, Phys. Rev. **D12**, 792 (1975)
83. M. Kurachi, R. Shrock, JHEP **0612**, 034 (2006) [arXiv:hep-ph/0605290]
84. R. Sundrum, S.D.H. Hsu, Nucl. Phys. B **391**, 127 (1993) [arXiv:hep-ph/9206225]
85. J.M. Maldacena, Adv. Theor. Math. Phys. **2**, 231 (1998) [Int. J. Theor. Phys. **38**, 1113 (1999)] [arXiv:hep-th/9711200]
86. D.K. Hong, H.U. Yee, Phys. Rev. **D74**, 015011 (2006) [arXiv:hep-ph/0602177]
87. J. Hirn, V. Sanz, Phys. Rev. Lett. **97**, 121803 (2006) [arXiv:hep-ph/0606086]
88. M. Piai, arXiv:hep-ph/0609104
89. K. Agashe, C. Csaki, C. Grojean, M. Reece, JHEP **0712**, 003 (2007) [arXiv:0704.1821 [hep-ph]]

90. C.D. Carone, J. Erlich, M. Sher, Phys. Rev. **D76**, 015015 (2007) [arXiv:0704.3084 [hep-th]]
91. T. Hirayama, K. Yoshioka, JHEP **0710**, 002 (2007) [arXiv:0705.3533 [hep-ph]]
92. D.D. Dietrich, M. Jarvinen, C. Kouvaris, arXiv:0908.4357 [hep-ph]
93. D.D. Dietrich, C. Kouvaris, Phys. Rev. **D79**, 075004 (2009) [arXiv:0809.1324 [hep-ph]]
94. D.D. Dietrich, C. Kouvaris, Constraining vectors and axial-vectors in walking technicolour by a holographic principle. Phys. Rev. **D78**, 055005 (2008) [arXiv:0805.1503 [hep-ph]]
95. C. Nunez, I. Papadimitriou, M. Piai, arXiv:0812.3655 [hep-th]
96. M. Fabbrichesi, M. Piai, L. Vecchi, Dynamical electro-weak symmetry breaking from deformed AdS: vector mesons and effective couplings. Phys. Rev. **D78**, 045009 (2008) [arXiv:0804.0124 [hep-ph]]
97. M. Kurachi, R. Shrock, K. Yamawaki, Phys. Rev. **D76**, 035003 (2007) [arXiv:0704.3481 [hep-ph]]
98. F. Sannino, Phys. Rev. **D82**, 081701 (2010) [arXiv:1006.0207 [hep-lat]]
99. F. Sannino, Phys. Rev. Lett. **105**, 232002 (2010) [arXiv:1007.0254 [hep-ph]]
100. S. Di Chiara, C. Pica, F. Sannino, Phys. Lett. B **700**, 229 (2011) [arXiv:1008.1267 [hep-ph]]

Chapter 2
Conformal Dynamics Interlude

Abstract We have seen that models of dynamical breaking of the electroweak symmetry are theoretically appealing and constitute one of the best motivated natural extensions of the standard model. These are also among the most challenging models to work with since they require deep knowledge of gauge dynamics in a regime where perturbation theory fails. In particular, it is of utmost importance to gain information on the nonperturbative dynamics of non-abelian four dimensional gauge theories. In this chapter we elucidate the physics of non-Abelian gauge theories as function of the gauge group, number of flavors, colors and matter representation.

2.1 Phases of Gauge Theories

Non-abelian gauge theories exist in a number of distinct phases which can be classified according to the characteristic dependence of the potential energy on the distance between two well separated static sources. The collection of all of these different behaviors, when represented, for example, in the flavor-color space, constitutes the *phase diagram* of the given gauge theory. The phase diagram of $SU(N)$ gauge theories as functions of number of the gauge group, flavors, colors and matter representation has been investigated in [1–14].

The analytical tools we will use for such an exploration are: (i) Precise results from higher order perturbation theory [14–16]; (ii) The conjectured all orders beta function for nonsupersymmetric gauge theories with fermionic matter in arbitrary representations of the gauge group [4, 6]; (ii) The truncated Schwinger-Dyson equation (SD) [17–19] (referred also as the ladder approximation in the literature).

We wish to study the phase diagram of any asymptotically free non-supersymmetric theories with fermionic matter transforming according to a generic representation of an SU(N) gauge group as function of the number of colors and flavors.

F. Sannino, *Dynamical Stabilization of the Fermi Scale*, SpringerBriefs in Physics,   31
DOI: 10.1007/978-3-642-33341-5_2, © The Author(s) 2013

We start by characterizing the possible phases via the potential $V(r)$ between two electric test charges separated by a large distance r. The list of possible potentials is given below:

$$\textbf{Coulomb}: \quad V(r) \propto \frac{1}{r} \tag{2.1}$$

$$\textbf{Free electric}: \quad V(r) \propto \frac{1}{r \log(r)} \tag{2.2}$$

$$\textbf{Free magnetic}: \quad V(r) \propto \frac{\log(r)}{r} \tag{2.3}$$

$$\textbf{Higgs}: \quad V(r) \propto \text{constant} \tag{2.4}$$

$$\textbf{Confining}: \quad V(r) \propto \sigma r. \tag{2.5}$$

A nice review of these phases can be found in [20] which here we reconsider for completeness. In the Coulomb phase, the electric charge $e^2(r)$ is a constant while in the free electric phase massless electrically charged fields renormalize the charge to zero at long distances as, i.e. $e^2(r) \sim 1/\log(r)$. QED is an abelian example of a free electric phase. The free magnetic phase occurs when massless magnetic monopoles renormalize the electric coupling constant at large distance with $e^2(r) \sim \log(r)$.

In the Higgs phase, the condensate of an electrically charged field gives a mass gap to the gauge fields by the Anderson-Higgs-Kibble mechanism and screens electric charges, leading to a potential which, up to an additive constant, has an exponential Yukawa decay to zero at long distances. In the confining phase, there is a mass gap with electric flux confined into a thin tube, leading to the linear potential with string tension σ.

We will be mainly interested in finding theories possessing a non-Abelian Coulomb phase or being close in the parameter space to these theories. In this phase we have massless interacting quarks and gluons exhibiting the Coulomb potential. This phase occurs when there is a non-trivial, infrared fixed point of the renormalization group. These are thus non-trivial, interacting, four dimensional conformal field theories. In the Coulomb phase the situation is actually more involved since for strong electrical charges the nonperturbative physical spectrum is much more involved.

To guess the behavior of the magnetic charge, at large distance separation, between two test magnetic charges one uses the Dirac condition:

$$e(r)g(r) \sim 1. \tag{2.6}$$

Then it becomes clear that $g(r)$ is constant in the Coulomb phase, increases with $\log(r)$ in the free electric phase and decreases as $1/\log(r)$ in the free magnetic phase. In these three phases the potential goes like $g^2(r)/r$. A linearly rising potential in

Turning Gauge Theory Knobs

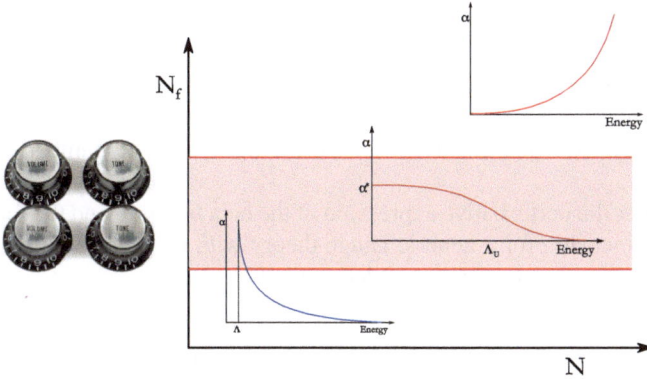

Fig. 2.1 A generic gauge theory has different Knobs one can tune. For example by changing the number of flavors one can enter in different phases. The *pink* region is the conformal region, i.e. the one where the coupling constant freezes at large distances (small energy). The region above the *pink* one corresponds to a non-abelian QED like theory and below to a QCD-like region. We have also plotted the cartoon of the running of the various coupling constants in the regions away from the boundaries of the conformal window. The *diagram* above is the qualitative one expected for a gauge theory with matter in the adjoint representation

the Higgs phase for magnetic test charges corresponds to the Meissner effect in the electric charges.

Confinement does not survive the presence of massless matter in the fundamental representation, such as light quarks in QCD. This is so since it is more convenient for the underlying theory to pop from the vacuum virtual quark-antiquark pairs when pulling two electric test charges apart. The potential for the confining phase will then change and there is no distinction between Higgs and confining phase.

Under electric-magnetic duality one exchanges electrically charged fields with magnetic ones then the behavior in the free electric phase is mapped in that of the free magnetic phase. The Higgs and confining phases are also expected to be exchanged under duality. Confinement can then be understood as the dual Meissner effect associated with a condensate of monopoles.

There is one more phase to consider which occurs when a given gauge theory at very high energies reaches an ultraviolet fixed point. The theory, in this case, is said to be asymptotically safe. Although there is not rigorous proof Weinberg [21] speculated that even the properly quantum corrected gravitational theory might develop such an ultraviolet fixed point which would *save* gravity from invalidating itself at and above the Planck scale. We refer to [22] for an up-to-date review and a list of relevant references on the subject (Fig. 2.1).

2.2 UV and IR Fixed Points of Gauge Theories at the Four Loops and Beyond

To gain a quantitative analytic understanding of the phase structure of different gauge theories we investigate the zeros of the perturbative beta function to the maximum known order and for one of the zeros also the limit of large number of flavors to all-orders.

We consider the perturbative expression of the beta function and the fermion mass anomalous dimension for a generic gauge theory with only fermionic matter in the $\overline{\text{MS}}$ scheme to four loops which was derived in [23, 24]:

$$\frac{da}{d \ln \mu^2} = \beta(a) = -\beta_0 a^2 - \beta_1 a^3 - \beta_2 a^4 - \beta_3 a^5 + O(a^6), \qquad (2.7)$$

$$-\frac{d \ln m}{d \ln \mu^2} = \frac{\gamma(a)}{2} = \gamma_0 a + \gamma_1 a^2 + \gamma_2 a^3 + \gamma_3 a^4 + O(a^5), \qquad (2.8)$$

where $m = m(\mu^2)$ is the renormalized (running) fermion mass and μ is the renormalization point in the $\overline{\text{MS}}$ scheme and $a = \alpha/4\pi = g^2/16\pi^2$ where $g = g(\mu^2)$ is the renormalized coupling constant of the theory.

The explicit expression of the coefficients above are reported in the Appendix A.2 for completeness. Note also that the beta function is gauge independent, order by order in perturbation theory [23]. The same also holds for the anomalous dimension of the fermion mass γ.

Here we report the investigation of the structure of the zeros of the four-loops beta function for any matter representation and gauge group [11, 15, 16]. Interestingly in [15] we found a *universal* classification of the behavior of the zeros as function of the number of flavors n_f.

To exemplify the various possible topologies emerging to this order, we plot the real nontrivial zeros as function of the number of flavors normalized to the one above which asymptotic freedom is lost ($\overline{n}_f$) in Fig. 2.2. The solid black lines represent the location of the ultraviolet (UV) zeros while the red-ones to the infrared (IR) stable fixed points, and finally the shaded areas are the regions where the β function is positive. To help visualizing the different regions the vertical axis is rescaled [15] according to the function $a^* = 2 \arctan(5a)/\pi$, mapping $[-\infty, +\infty]$ into the interval $[-1, 1]$.

A straight vertical line corresponds to a fixed value of n_f and the intersection of this line with the solid curves determines the number of the zeros, the color of the curves the type of zeros (if red is IR and if black is UV), and finally the corresponding horizontal value is the coupling location. The landscape of the zeros was termed *zerology* in [15].

We investigated also the negative values of α since this is the most natural mathematical setting. In fact, the properties of the pure Yang-Mills theory at negative

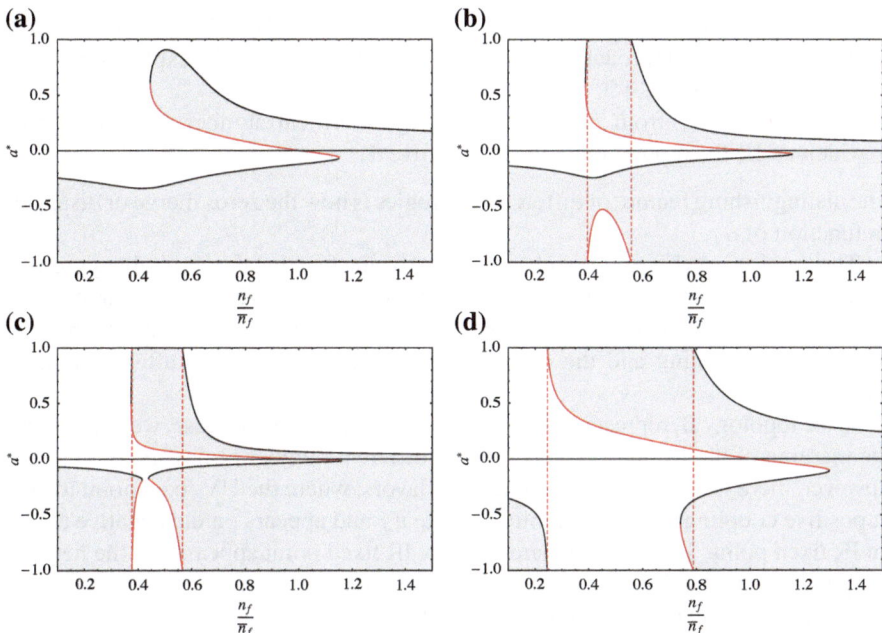

Fig. 2.2 The four different topologies displayed above classify the entire *zerology* landscape. We show, in each plot, the regions of positive (*gray*) and negative (*white*) values of the beta function for different gauge theories. The *solid lines*, per each figure, are the locations of the zeros of the beta functions. The *lines* of UV fixed points are in *black* while the IR ones in *red*. We have defined $a^* = \dfrac{2}{\pi} \arctan(5a)$. The *vertical dashed red-lines* correspond to the location where one zero approaches infinity: **a** first kind of topology, **b** second kind of topology, **c** third kind of topology, **d** fourth kind of topology

α were studied on the lattice by Li and Meurice in [25] showing interesting relations between the positive and negative regions of α.

By explicit enumeration [15] it is possible to identify just four distinct topologies covering the full available zerology for any gauge group and matter representation reported in Fig. 2.2.

The four-loop analysis tells us that [15]:

- At small number of flavors there is only a negative ultraviolet zero.
- At around and above $\bar{n}_f$ we observe the existence of three zeros, two ultraviolets and one infrared. The infrared one, near $\bar{n}_f$, is the Banks-Zaks [26] point. Above $\bar{n}_f$, the IR fixed point is now at a negative value of α and at a new critical number of flavors collides with the UV fixed point zero at a negative value of the coupling, forming a double zero. At this point the beta function is positive for any negative alpha.

- At very large number of flavors the UV fixed point, for positive values of alpha, always exists and approaches zero asymptotically as $n_f^{-2/3}$. The explicit derivation is provided in Sect. 2.4.3.
- By increasing n_f from zero there is always a critical number of flavors above which an IR fixed point emerges for positive α.

The distinguishing feature of different topologies is how the zeros merge or disappear as function of n_f.

The topology A (Fig. 2.2a) is characterized by the fact that the zeros always remain at finite values of the coupling. This means that when a zero disappear it has to annihilate with another one. This happens at two distinct locations. One at a positive value of the coupling and the other at a negative one occurring after asymptotic freedom is lost.

In the topology B, represented in Fig. 2.2b, as for the previous case, we still observe the merging of the IR and UV zeros at two different number of flavors. In this case, however, there is a region in the number of flavors, where the UV fixed point located at positive couplings reaches infinity at finite n_f and appears on the negative axis as an IR fixed point. The region where the new IR fixed point appears (on the negative coupling constant axis) ends before asymptotic freedom is lost.

The defining feature shown in Fig. 2.2c for topology C is that the appearance of two more merging points at negative values of α.

In Fig. 2.2d, topology D, one observes that the IR zero at a positive value of the coupling reachers infinity at a finite value of the number of flavors, which is the distinctive feature of this topology.

A new feature at the four-loop order is that two positive nontrivial zeros, one IR and the other UV, can emerge simultaneously and can annihilate at a particular value of n_f. At the two-loop level this feature does not exist and, in particular, no nontrivial ultraviolet fixed point is seen.

As an example where these topologies arise we consider $SU(N)$ with fundamental fermions as function of N. For $N = 2$ and 3 the topology A occurs. Increasing N the maximum value reached by the positive UV zero increases and for $N = 4$ it reaches infinity and therefore it enters topology B. Increasing N further the local maximum of the IR negative zero-curve increases till it pinches the UV negative zero line for $N = 11$ entering topology C. Topology D is not realized in this case. On the other hand any $SU(N)$ gauge theory with $N \geq 2$ fermions and fermions in the adjoint representation lead to topology D.

In Table 2.1 we provide a catalogue of the four-loop zerology for $SU(N)$, $SO(N)$ and $SP(2N)$ gauge theories with fermions transforming according to the fundamental and the 2-index representations.

Table 2.1 Catalogue of the four-loop zerology for $SU(N)$, $SO(N)$ and $SP(2N)$ gauge theories with fermions transforming according to the fundamental and the 2-index representations

Rep.	Top. A	Top. B	Top. C	Top. D
$SU(N)$				
FUND	$N = 2, 3$	$4 \leq N \leq 11$	$N \geq 12$	–
ADJ	–	–	–	$N \geq 2$
2-SYM	–	–	–	$N \geq 2$
2-ASY	$N = 3, 4, 5$	$N = 6, 7$	$8 \leq N \leq 26$	$N \geq 27$
$SO(N)$				
FUND	–	$N \leq 6$	$N = 5$	$N = 3, 4$
ADJ	–	–	–	$N \geq 3$
2-SYM	–	–	–	$N \geq 3$
$SP(2N)$				
FUND	$N = 1, 2$	$3 \leq N \leq 4$	$N \geq 5$	–
ADJ	–	–	–	$N \geq 1$
2-ASY	$N = 3, 4$	$N = 2, 5$	$6 \leq N \leq 14$	$N \geq 15$

2.3 Conformal Window

The conformal window is defined as the region in theory space, as function of number of flavors and colors where the underlying gauge theory displays large distance conformality for a positive value of the coupling α. $\overline{n}_f$ constitutes the upper boundary of the conformal window and the lower boundary here is estimated by identifying for which number of flavors the theory looses the infrared fixed point at a given number of colors. Several methods have been used to constrain the conformal window for non-supersymmetric gauge theories. We will review only a few here which has been proven to be useful either to guide lattice simulations or that have been shown effective in producing relevant physical quantities.

2.4 Four-Loop Conformal Window

Having at our disposal the four-loops beta function we use it to estimate the lower boundary of the conformal window. However, due to the fact that it is obviously a truncated beta function the true window can be quantitatively different.

The results for the $SU(N)$ gauge groups are presented in Fig. 2.3 for the fundamental, two-index symmetric, two-index antisymmetric and adjoint representation. The conformal window at the four-loop level is considerably wider, for any representation, when compared with the Schwinger-Dyson results [1, 2] or the one obtained using the critical number of flavors where the free energy changes sign, as suggested in [11]. For completeness the conformal window for the orthogonal and symplectic

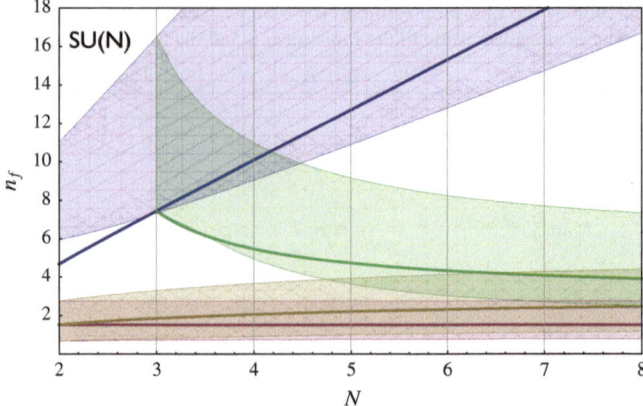

Fig. 2.3 Conformal window for $SU(N)$ groups for the fundamental representation (*upper light-blue*), two-index antisymmetric (next to the *highest light-green*), two-index symmetric (third window from the *top light-brown*) and finally the adjoint representation (*bottom light-pink*). The lower boundary corresponds to the point where the infrared fixed point disappears at four loops. The *solid thick lines* correspond to the number of flavors for which the all-orders beta function predicts an anomalous dimension equal to unity

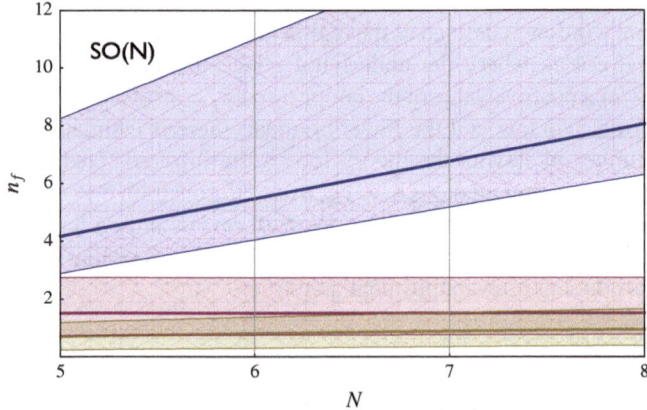

Fig. 2.4 Conformal window for $SO(N)$ groups for the fundamental representation (*upper light-blue*), two-index antisymmetric (which is the adjoint and second from the top (*pink*-region)), two-index symmetric (bottom window in *light-brown*)

gauge groups is also shown respectively in Figs. 2.4 and 2.5. There is a universal trend towards the widening of the conformal regions with respect to earlier estimates using other nonperturbative methods.

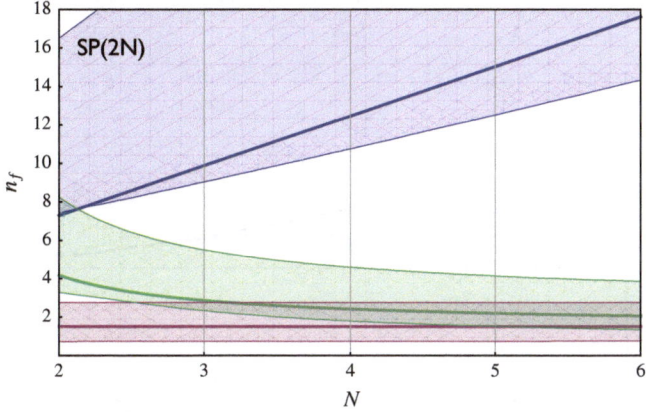

Fig. 2.5 Conformal window for $SP(2N)$ groups for the fundamental representation (*upper light-blue*), two-index antisymmetric (next to the *highest light-green*), two-index symmetric, i.e. the adjoint, (bottom window in *light-pink*)

2.4.1 All-Orders Beta Function Comparison

We have recently argued for the existence of a scheme in which an all-orders beta function [6] assumes the form:

$$\frac{\beta(a)}{a} = -\frac{a}{3} \frac{11 C_2(G) - 2T(r) n_f (2 + \Delta_F \gamma)}{1 - 2a \frac{17}{11} C_2(G)}, \qquad (2.9)$$

with

$$\Delta_F = 1 + \frac{7}{11} \frac{C_2(G)}{C_2(r)}. \qquad (2.10)$$

The group invariants defined in Appendix A.2. The *scheme independent* analytical expression of the anomalous dimension of the mass at the IR positive zero is:

$$\gamma = \frac{11 C_2(G) - 4T(r) n_f}{2 n_f T(r) \left(1 + \frac{7}{11} \frac{C_2(G)}{C_2(r)}\right)}. \qquad (2.11)$$

We plot, for reference, in Figs. 2.3, 2.4 and 2.5 the lines corresponding to this anomalous dimension equal to unity. These are the solid thick curves for the different representations. These lines could be viewed as the lower boundary of the conformal window if it is marked by the anomalous dimension to be unity. The size of these regions are consistent with the ones derived via gauge dualities in [27, 28].

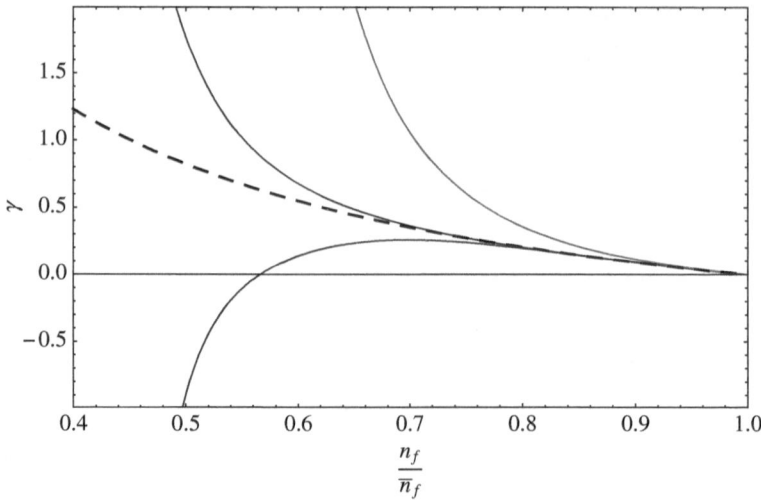

Fig. 2.6 Anomalous dimension of the mass, at the infrared fixed point, for $SU(3)$ as function of the number of fundamental flavors at two loops (*upper brown-curve*), three-loops (second *curve* from the top in *magenta*), all-orders (*dashed-curve* in *black*), four-loops (*bottom curve* in *blue*)

2.4.2 Four-Loop Anomalous Dimensions

In Fig. 2.6 we plot the anomalous dimension of the mass for the $SU(3)$ gauge theory, as function of the number of fundamental flavors, at the IR fixed point as derived in [15, 16]. The three solid lines correspond respectively, from top to bottom, to the two-, three- and four-loop results. Of course, perturbation theory is reliable only in a small range of flavors near $\overline{n}_f$. A similar behavior is observed for any other gauge group, matter representation and different number of colors.

Having at hand an all-order scheme-independent result, we compare it with the perturbative one. The dashed line, in Fig. 2.6, is the all-order anomalous dimension from Eq. (2.11). It is striking that the all-order result is much more well behaved than the four-loop predictions which, in this example, reach large and negative values long before loosing the IR positive zero.

Due to the phenomenological interest in models of minimal walking technicolor [1, 29, 30] we report the anomalous dimension at the fixed point also for the $SU(2)$ gauge theory with two-adjoint fermions in Fig. 2.7. These theories are being subject to intensive numerical investigations via lattice simulations [31–59].

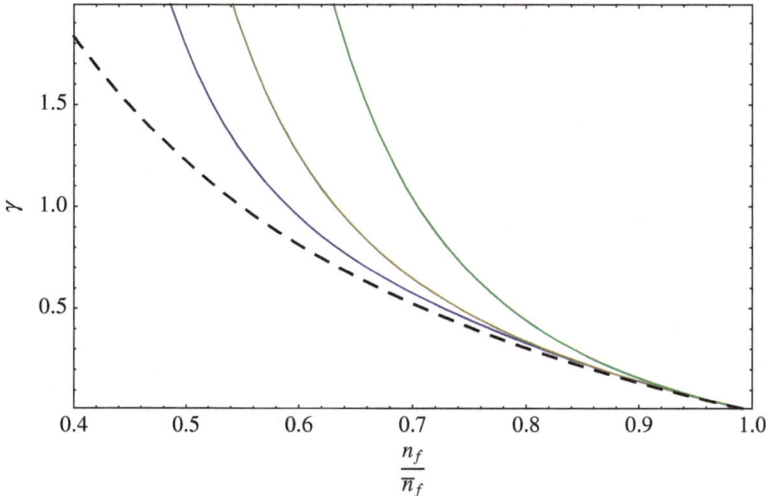

Fig. 2.7 Anomalous dimension of the mass, at the infrared fixed point, for $SU(2)$ as function of the number of adjoint Dirac flavors at two loops (*up green-curve*), three-loops (second *curve* from the *top*), four-loops (third *curve* in *blue*), all-orders (*dashed curve* in *black*)

2.4.3 Asymptotic Safety at Large n_f: A New Phase

To the four-loop order a positive UV zero appears for a sufficiently large number of flavors. We observed that the value of the zero as function of number of flavors decreases monotonically as $n_f^{-2/3}$ at four loops. In fact, it is possible to generalize this behavior to any finite order in perturbation theory. Consider the equation for the zeros of the beta function in which the leading powers in the number of flavors are made explicit:

$$b_0 n_f + \sum_{k=1}^{\infty} b_k \, n_f^k \, \alpha^k = 0 \,, \tag{2.12}$$

where $b_0 = \beta_0/n_f$ and $b_k = \beta_k/n_f^k$. We used the fact that the first and second coefficient of the beta function are linear in the number of flavors and, in general, the successive coefficients have one extra power of n_f [60]. Therefore the coefficients b_k are finite at large number of flavors.

We define:

$$x = n_f \alpha \,, \tag{2.13}$$

and the equation at any fixed perturbative order P reads:

$$b_0 n_f + \sum_{k=1}^{P} b_k \, x^k = 0 \,. \tag{2.14}$$

At large n_f the solution approaches:

$$x = \left(-\frac{b_0 n_f}{b_P}\right)^{\frac{1}{P}} \longrightarrow \alpha = \left(-\frac{b_0}{b_P}\right)^{\frac{1}{P}} n_f^{\frac{1-P}{P}} . \tag{2.15}$$

There are P complex solutions for x lying on a circle in the complex plane. A positive solution exists only if b_P is positive at large n_f. This is indeed the case, at the four-loop order, for any gauge theory showing that the UV positive zero vanishes as $n_f^{-2/3}$. If this UV zero persists to higher orders its location will change albeit will vanish faster as a function of n_f when increasing P, i.e. the exponent $(1 - P)/P$ increases in absolute value. The case $n_f^{-2/3}$ is recovered for $P = 3$.

Interestingly it is possible to sum exactly the perturbative infinite sum for the beta function, at large of number of flavors given that the leading coefficients are known. The result is:

$$\frac{3}{4n_f T_F} \frac{\beta(a)}{a^2} = 1 + \frac{H(x)}{n_f} + \mathcal{O}\left(n_f^{-2}\right). \tag{2.16}$$

The explicit form of $H(x)$ can be found in [60]. The important feature, here, is that $H(x)$ possesses a negative singularity at $x = 3\pi/T_F$. This demonstrates that there always is a solution for the existence of a nontrivial UV fixed point at the leading order in n_f for the following positive value of the coupling:

$$\alpha_{UV} = \frac{3\pi}{T_F n_f} . \tag{2.17}$$

The function $H(x)$ has also other singularities which might signal the presence of new zeros which we will not consider here, but that would be worth exploring.

Higher order terms in n_f^{-1} can, in principle, modify the result if the singularity structure is such to remove or modify its location.

A more complete discussion of the singularity structure of the coefficients of the n_f^{-1} expansion has appeared in [60] also for QED. It seems plausible that the smallest UV fixed point is an all-orders feature.

2.4.4 Schwinger-Dyson in the Rainbow Approximation

For nonsupersymmetric theories an old way to get quantitative estimates is to use the *rainbow* approximation to the Schwinger-Dyson equation [61, 62], see Fig. 2.8. Here the full nonperturbative fermion propagator in momentum space reads

$$iS^{-1}(p) = Z(p) \left(\not{p} - \Sigma(p)\right) , \tag{2.18}$$

and the Euclidianized gap equation in Landau gauge is given by

Fig. 2.8 Rainbow approximation for the fermion self energy function. The boson is a gluon

$$\Sigma(p) = 3C_2(r) \int \frac{d^4k}{(2\pi)^4} \frac{\alpha\left((k-p)^2\right)}{(k-p)^2} \frac{\Sigma(k^2)}{Z(k^2)k^2 + \Sigma^2(k^2)}, \qquad (2.19)$$

where $Z(k^2) = 1$ in the Landau gauge and we linearize the equation by neglecting $\Sigma^2(k^2)$ in the denominator. Upon converting it into a differential equation and assuming that the coupling $\alpha(\mu) \approx \alpha_c$ is varying slowly ($\beta(\alpha) \simeq 0$) one gets the approximate (WKB) solutions

$$\Sigma(p) \propto p^{-\gamma(\mu)}, \qquad \Sigma(p) \propto p^{\gamma(\mu)-2}. \qquad (2.20)$$

The critical coupling is given in terms of the quadratic Casimir of the representation of the fermions

$$\alpha_c \equiv \frac{\pi}{3C_2(r)}. \qquad (2.21)$$

The anomalous dimension of the fermion mass operator is

$$\gamma(\mu) = 1 - \sqrt{1 - \frac{\alpha(\mu)}{\alpha_c}} \sim \frac{3C_2(r)\alpha(\mu)}{2\pi}. \qquad (2.22)$$

The first solution corresponds to the running of an ordinary mass term (*hard* mass) of nondynamical origin and the second solution to a *soft* mass dynamically generated. In fact in the second case one observes the $1/p^2$ behavior in the limit of large momentum.

Within this approximation spontaneous symmetry breaking occurs when α reaches the critical coupling α_c given in Eq. (2.21). From Eq. (2.22) it is clear that α_c is reached when γ is of order unity [17, 18, 63]. Hence the symmetry breaking occurs when the soft and the hard mass terms scale as function of the energy scale in the same way. In Ref. [17], it was noted that in the lowest (ladder) order, the gap equation leads to the condition $\gamma(2 - \gamma) = 1$ for chiral symmetry breaking to occur. To all orders in perturbation theory this condition is gauge invariant and also equivalent nonperturbatively to the condition $\gamma = 1$. However, to any finite order in perturbation theory these conditions are, of course, different. Interestingly the condition $\gamma(2 - \gamma) = 1$ leads again to the critical coupling α_c when using the perturbative leading order expression for the anomalous dimension which is $\gamma = \frac{3C_2(r)}{2\pi}\alpha$.

To summarize, the idea behind this method is simple. One simply compares the two couplings in the infrared associated to (i) an infrared zero in the β function, call it α^* with (ii) the critical coupling, denoted with α_c, above which a dynamical mass for the fermions generates nonperturbatively and chiral symmetry breaking occurs. If α^*

is less than α_c chiral symmetry does not occur and the theory remains conformal in the infrared, viceversa if α^* is larger than α_c then the fermions acquire a dynamical mass and the theory cannot be conformal in the infrared. The condition $\alpha^* = \alpha_c$ provides the desired n_f^{SD} as function of N. In practice to estimate α^* one uses the two-loop beta function while the truncated SD equation to determine α_c as we have done before. This corresponds to when the anomalous dimension of the quark mass operator becomes approximately unity.

The two-loop fixed point value of the coupling constant is:

$$\frac{\alpha^*}{4\pi} = -\frac{\beta_0}{\beta_1}. \tag{2.23}$$

with the following definition of the two-loop beta function

$$\beta(g) = -\frac{\beta_0}{(4\pi)^2}g^3 - \frac{\beta_1}{(4\pi)^4}g^5, \tag{2.24}$$

where g is the gauge coupling and the beta function coefficients are given by

$$\beta_0 = \frac{11}{3}C_2(G) - \frac{4}{3}T(r)n_f \tag{2.25}$$

$$\beta_1 = \frac{34}{3}C_2^2(G) - \frac{20}{3}C_2(G)T(r)n_f - 4C_2(r)T(r)n_f. \tag{2.26}$$

To this order the two coefficients are universal, i.e. do not depend on which renormalization group scheme one has used to determine them. The perturbative expression for the anomalous dimension reads:

$$\gamma(g^2) = \frac{3}{2}C_2(r)\frac{g^2}{4\pi^2} + O(g^4). \tag{2.27}$$

with $\gamma = -d\ln m/d\ln\mu$ and m the renormalized fermion mass.

For a fixed number of colors the critical number of flavors for which the order of α^* and α_c changes is defined by imposing $\alpha^* = \alpha_c$, and it is given by

$$n_f^{\text{SD}} = \frac{17C_2(G) + 66C_2(r)}{10C_2(G) + 30C_2(r)}\frac{C_2(G)}{T(r)}. \tag{2.28}$$

Comparing with the previous results obtained using the four-loop approximation or the conjectured all orders beta function the striking differences is that the anomalous dimension of the fermion mass operator, for the same number of flavors, is typically overestimated by the SD analysis while the conformal window is smaller, i.e. critical number of flavors is predicted to be higher than other methods.

2.5 Gauge Duals and Conformal Window

One of the most fascinating possibilities is that generic asymptotically free gauge theories have magnetic duals. In fact, in the late nineties, in a series of ground breaking papers Seiberg [64, 65] provided strong support for the existence of a consistent picture of such a duality within a supersymmetric framework. Supersymmetry is, however, quite special and the existence of such a duality does not automatically imply the existence of nonsupersymmetric duals. One of the most relevant results put forward by Seiberg has been the identification of the boundary of the conformal window for supersymmetric QCD as function of the number of flavors and colors. The dual theories proposed by Seiberg pass a set of mathematical consistency relations known as 't Hooft anomaly conditions (in [66]). Another important tool has been the knowledge of the all orders supersymmetric beta function [67–69].

Arguably the existence of a possible dual of a generic nonsupersymmetric asymptotically free gauge theory able to reproduce its infrared dynamics must match the 't Hooft anomaly conditions [66].

We have exhibited several solutions of these conditions for QCD in [27] and for certain gauge theories with higher dimensional representations in [28]. An earlier exploration already appeared in the literature [70]. The novelty with respect to these earlier results are: (i) The request that the gauge singlet operators associated to the magnetic baryons should be interpreted as bound states of ordinary baryons [27]; (ii) The fact that the asymptotically free condition for the dual theory matches the lower bound on the conformal window obtained using the all orders beta function [4]. These extra constraints help restricting further the number of possible gauge duals without diminishing the exactness of the associate solutions with respect to the 't Hooft anomaly conditions.

We will briefly summarize here the novel solutions to the 't Hooft anomaly conditions for QCD and the theories with higher dimensional representations. The resulting *magnetic* dual allows to predict the critical number of flavors above which the asymptotically free theory, in the electric variables, enters the conformal regime as predicted using the all orders conjectured beta function [4].

2.5.1 QCD Duals

The underlying gauge group is $SU(3)$ while the quantum flavor group is

$$SU_L(n_f) \times SU_R(n_f) \times U_V(1) , \qquad (2.29)$$

and the classical $U_A(1)$ symmetry is destroyed at the quantum level by the Adler-Bell-Jackiw anomaly. We indicate with $Q^i_{\alpha;c}$ the two component left spinor where $\alpha = 1, 2$ is the spinor index, $c = 1, \ldots, 3$ is the color index while $i = 1, \ldots, n_f$ represents

Table 2.2 Field content of an SU(3) gauge theory with quantum global symmetry $SU_L(n_f) \times SU_R(n_f) \times U_V(1)$

Fields	$[SU(3)]$	$SU_L(n_f)$	$SU_R(n_f)$	$U_V(1)$
Q	□	□	1	1
$\tilde{Q}$	$\overline{\square}$	1	$\overline{\square}$	−1
G_μ	Adj	1	1	1

't Hooft Anomaly Matching

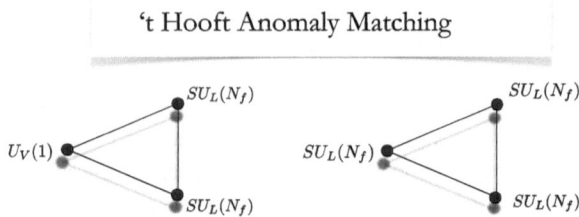

Fig. 2.9 The 't Hooft anomaly matching conditions are related to the saturation of the global anomalies stemming out of the one-loop triangle diagrams represented, for the theory of interest, here. According to 't Hooft both theories, i.e. the electric and the magnetic ones, should yield the same global anomalies

the flavor. $\tilde{Q}_i^{\alpha;c}$ is the two component conjugated right spinor. We summarize the transformation properties in Table 2.2.

The global anomalies are associated to the triangle diagrams featuring at the vertices three $SU(n_f)$ generators (either all right or all left), or two $SU(n_f)$ generators (all right or all left) and one $U_V(1)$ charge. We indicate these anomalies for short with:

$$SU_{L/R}(n_f)^3, \qquad SU_{L/R}(n_f)^2\, U_V(1). \tag{2.30}$$

For a vector like theory there are no further global anomalies (Fig. 2.9). The cubic anomaly factor, for fermions in fundamental representations, is 1 for Q and −1 for $\tilde{Q}$ while the quadratic anomaly factor is 1 for both leading to

$$SU_{L/R}(n_f)^3 \propto \pm 3, \quad SU_{L/R}(n_f)^2 U_V(1) \propto \pm 3. \tag{2.31}$$

If a magnetic dual of QCD does exist one expects it to be weakly coupled near the critical number of flavors below which one breaks large distance conformality in the electric variables. This idea is depicted in Fig 2.10.

Determining a possible unique dual theory for QCD is, however, not simple given the few mathematical constraints at our disposal, as already observed in [70]. The saturation of the global anomalies is an important tool but is not able to select out a unique solution. We shall see, however, that one of the solutions, when interpreted as the QCD dual, leads to a prediction of a critical number of flavors corresponding exactly to the one obtained via the conjectured all orders beta function.

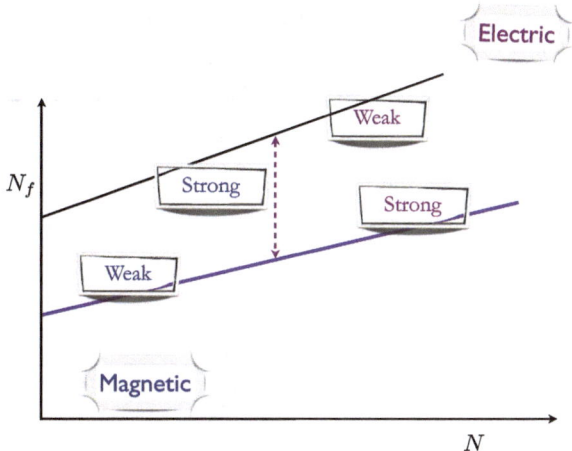

Fig. 2.10 Schematic representation of the phase diagram as function of number of flavors and colors. For a given number of colors by increasing the number flavors within the conformal window we move from the *lowest line* (*violet*) to the *upper* (*black*) one. The *upper black line* corresponds to the one where one looses asymptotic freedom in the electric variables and the *lower line* where chiral symmetry breaks and long distance conformality is lost. In the *magnetic* variables the situation is reverted and the perturbative line, i.e. the one where one looses asymptotic freedom in the magnetic variables, correspond to the one where chiral symmetry breaks in the electric ones

We seek solutions of the anomaly matching conditions for a gauge theory $SU(X)$ with global symmetry group $SU_L(n_f) \times SU_R(n_f) \times U_V(1)$ featuring *magnetic* quarks q and $\tilde{q}$ together with $SU(X)$ gauge singlet states identifiable as baryons built out of the *electric* quarks Q. Since mesons do not affect directly global anomaly matching conditions we could add them to the spectrum of the dual theory. We study the case in which X is a linear combination of number of flavors and colors of the type $\alpha n_f + 3\beta$ with α and β integer numbers.

We add to the *magnetic* quarks gauge singlet Weyl fermions which can be identified with the baryons of QCD but massless. The generic dual spectrum is summarized in Table 2.3.

The wave functions for the gauge singlet fields A, C and S are obtained by projecting the flavor indices of the following operator

$$\varepsilon^{c_1 c_2 c_3} Q_{c_1}^{i_1} Q_{c_2}^{i_2} Q_{c_3}^{i_3}, \tag{2.32}$$

over the three irreducible representations of $SU_L(n_f)$ as indicated in the Table 2.3. These states are all singlets under the $SU_R(n_f)$ flavor group. Similarly one can construct the only right-transforming baryons $\tilde{A}$, $\tilde{C}$ and $\tilde{S}$ via $\tilde{Q}$. The B states are made by two Q fields and one right field $\overline{\tilde{Q}}$ while the D fields are made by one Q and two $\overline{\tilde{Q}}$ fermions. y is the, yet to be determined, baryon charge of the *magnetic* quarks while the baryon charge of composite states is fixed in units of the QCD quark one.

Table 2.3 Massless spectrum of *magnetic* quarks and baryons and their transformation properties under the global symmetry group

Fields	$[SU(X)]$	$SU_L(n_f)$	$SU_R(n_f)$	$U_V(1)$	# of copies
q	$\square$	$\square$	1	y	1
$\tilde{q}$	$\bar{\square}$	1	$\bar{\square}$	$-y$	1
A	1	(antisym. tableau)	1	3	ℓ_A
S	1	(sym. tableau)	1	3	ℓ_S
C	1	(mixed tableau)	1	3	ℓ_C
B_A	1	(antisym. tableau)	$\square$	3	ℓ_{B_A}
B_S	1	(sym. tableau)	$\square$	3	ℓ_{B_S}
D_A	1	$\square$	(antisym. tableau)	3	ℓ_{D_A}
D_S	1	$\square$	(sym. tableau)	3	ℓ_{D_S}
$\tilde{A}$	1	1	(antisym. tableau)	-3	$\ell_{\tilde{A}}$
$\tilde{S}$	1	1	(sym. tableau)	-3	$\ell_{\tilde{S}}$
$\tilde{C}$	1	1	(mixed tableau)	-3	$\ell_{\tilde{C}}$

The last column represents the multiplicity of each state and each state is a Weyl fermion

The ℓs count the number of times the same baryonic matter representation appears as part of the spectrum of the theory. Invariance under parity and charge conjugation of the underlying theory requires $\ell_J = \ell_{\tilde{J}}$ with $J = A, S, \ldots, C$ and $\ell_B = -\ell_D$.

Having defined the possible massless matter content of the gauge theory dual to QCD we compute the $SU_L(n_f)^3$ and $SU_L(n_f)^2\, U_{V(1)}$ global anomalies in terms of the new fields:

$$SU_L(n_f)^3 \propto X + \frac{(n_f-3)(n_f-6)}{2}\,\ell_A$$
$$+ \frac{(n_f+3)(n_f+6)}{2}\,\ell_S + (n_f^2-9)\,\ell_C$$
$$+ (n_f-4)n_f\,\ell_{B_A} + (n_f+4)n_f\,\ell_{B_S} + \frac{n_f(n_f-1)}{2}\,\ell_{D_A}$$
$$+ \frac{n_f(n_f+1)}{2}\,\ell_{D_S} = 3\,, \tag{2.33}$$

$$SU_L(n_f)^2\, U_V(1) \propto y\,X + 3\frac{(n_f-3)(n_f-2)}{2}\,\ell_A$$
$$+ 3\frac{(n_f+3)(n_f+2)}{2}\,\ell_S + 3(n_f^2-3)\,\ell_C$$
$$+ 3(n_f-2)n_f\,\ell_{B_A} + 3(n_f+2)n_f\,\ell_{B_S} + 3\frac{n_f(n_f-1)}{2}\,\ell_{D_A}$$
$$+ 3\frac{n_f(n_f+1)}{2}\,\ell_{D_S} = 3\,. \tag{2.34}$$

The right-hand side is the corresponding value of the anomaly for QCD.

Table 2.4 Massless spectrum of *magnetic* quarks and baryons and their transformation properties under the global symmetry group

Fields	$\left[SU(2n_f - 5N)\right]$	$SU_L(n_f)$	$SU_R(n_f)$	$U_V(1)$	# of copies
q	☐	☐	1	$\dfrac{N(2n_f-5)}{2n_f-5N}$	1
$\tilde{q}$	$\overline{\square}$	1	$\overline{\square}$	$-\dfrac{N(2n_f-5)}{2n_f-5N}$	1
A	1	⊟	1	3	2
B_A	1	⊟	☐	3	−2
D_A	1	☐	⊟	3	2
$\tilde{A}$	1	1	$\overline{\boxminus}$	−3	2

The last column represents the multiplicity of each state and each state is a Weyl fermion

2.5.2 A Realistic QCD Dual

We have found several solutions to the anomaly matching conditions presented above. Some were found previously in [70]. Here we start with a new solution in which the gauge group is $SU(2n_f - 5N)$ with the number of colors N equal to 3. It is, however, convenient to keep the dependence on N explicit.

The solution above corresponds to the following value assumed by the indices and y baryonic charge in Table 2.4.

$$X = 2n_f - 5N, \quad \ell_A = 2, \quad \ell_{D_A} = -\ell_{B_A} = 2,$$
$$\ell_S = \ell_{B_S} = \ell_{D_S} = \ell_C = 0, \quad y = N\frac{2n_f - 5}{2n_f - 15}, \tag{2.35}$$

with $N = 3$. X must assume a value strictly larger than one otherwise it is an abelian gauge theory. This provides the first nontrivial bound on the number of flavors:

$$n_f > \frac{5N + 1}{2}, \tag{2.36}$$

which for $N = 3$ requires $n_f > 8$.

2.5.3 Conformal Window from the Dual Magnetic Theory

Asymptotic freedom of the newly found theory is dictated by the coefficient of the one-loop beta function:

$$\beta_0 = \frac{11}{3}(2n_f - 5N) - \frac{2}{3}n_f. \tag{2.37}$$

To this order in perturbation theory the gauge singlet states do not affect the magnetic quark sector and we can hence determine the number of flavors obtained by requiring the dual theory to be asymptotic free. i.e.:

$$n_f \geq \frac{11}{4}N = 2.75 \qquad\qquad \text{Dual Asymptotic Freedom}. \tag{2.38}$$

Quite remarkably this value is close to the one predicted by means of the all orders conjectured beta function for the lowest bound of the conformal window, in the *electric* variables, when taking the anomalous dimension of the mass to be $\gamma = 1$. I.e. at large N[1]

$$n_f^{BF}|_{\gamma=1} \simeq 2.57N. \tag{2.39}$$

For $N = 3$ duality would seem to require the critical number of flavors to be 8.25.[2] We consider this a nontrivial and interesting result.

To investigate the decoupling of each flavor at the time one needs to introduce bosonic degrees of freedom. These are not constrained by anomaly matching conditions. Interactions among the mesonic degrees of freedom and the fermions in the dual theory cannot be neglected in the regime when the dynamics is strong. The simplest mesonic operator M_i^j transforming simultaneously according to the antifundamental representation of $SU_L(n_f)$ and the fundamental representation of $SU_R(n_f)$ leads to the following type of interactions for the dual theory:

$$\begin{aligned}
L_M = \ & Y_{q\bar{q}}\, q\, M\, \tilde{q} + Y_{AB_A}\, AM\overline{B}_A + Y_{CB_A}\, CM\overline{B}_A \\
& + Y_{CB_S}\, C\, M\overline{B}_S + Y_{SB_S}\, SM\overline{B}_S \\
& + Y_{B_A D_A}B_A M\overline{D}_A + Y_{B_A D_S}\, B_A\, M\overline{D}_S \\
& + Y_{B_S D_A}B_S M\overline{D}_A + Y_{B_S D_S}B_S M\overline{D}_S + \text{h.c.}
\end{aligned} \tag{2.40}$$

The coefficients of the various operators are matrices taking into account the multiplicity with which each state occurs. The number of operators drastically reduces if we consider only the ones linear in M. The dual quarks and baryons interact via mesonic exchanges. We have considered only the meson field for the bosonic spectrum because it is the one with the most obvious interpretation in terms on the electric variables. One can also envision adding new scalars charged under the dual gauge group [70] and in this case one can have contact interactions between the magnetic quarks and baryons. We expect these operators to play a role near the lower bound of the conformal window of the magnetic theory where QCD is expected to become free. It is straightforward to adapt the terms above to any anomaly matching solution.

---

[1] This result differs from the one found in the original paper [27] and reported in the earlier review [71] because, in the meanwhile, the all-orders beta function was corrected in [6].

[2] Actually given that X must be at least 2 we must have $n_f \geq 8.5$ rather than 8.25.

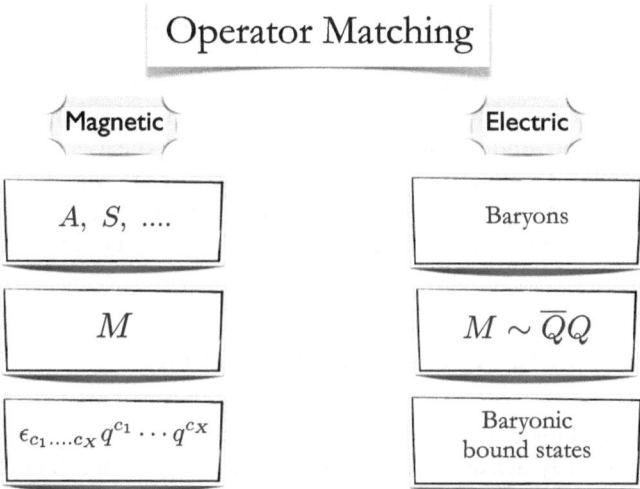

Fig. 2.11 We propose the above correspondence between the gauge singlet operators of the magnetic theory and the electric ones. The novelty introduced in [27] with respect to any of the earlier approaches is the identification of the *magnetic* baryons, i.e. the ones constructed via the magnetic quarks, with bound states of baryons in the electric variables

In Seiberg's analysis it was also possible to match some of the operators of the magnetic theory with the ones of the electric theory. The situation for QCD is, in principle, more involved although it is clear that certain magnetic operators match exactly the respective ones in the electric variables. These are the meson M and the massless baryons, $A, \widetilde{A}, \ldots, S$ shown in Table 2.3. The baryonic type operators constructed via the magnetic dual quarks have baryonic charge which is a multiple of the ordinary baryons and, hence, we propose to identify them, in the electric variables, with bound states of QCD baryons. We summarize the proposed operator matching constraints in Fig. 2.11.

The generalization to a generic number of colors is currently under investigation. It is an interesting issue and to address it requires the knowledge of the spectrum of baryons for arbitrary number of colors. It is reasonable to expect, however, a possible nontrivial generalization to any number of odd colors[3]. A relevant application of gauge duality has been to determine the left-right vector two-point function correlator at the lower boundary of the conformal window [72].

2.6 Walking Versus Jumping Dynamics

We concentrated our efforts mostly on the size of the conformal window neglecting, almost entirely, what happens at the boundary between the conformally intact phase and the broken one. However, despite much efforts we do not yet know the physical

---

[3] For an even number of colors the baryons are bosons and a the analysis must modify.

properties at the boundary between a conformally broken and a conformally restored phase for generic gauge theories. This problem remains an important mystery to solve. A famous conjecture has been put forward some time ago [19, 73] and it is known as Miransky scaling. In [74] it was argued for the potential existence of another intriguing possibility leading to a radically different near-conformal behavior. For setting up the stage we start with a brief review of the Miransky scaling and modeling. Following [74] we then introduce the alternative scenario and deduce the consequences for models of dynamical electroweak symmetry breaking.

2.6.1 Miransky Scaling and Walking Dynamics

This scaling arises under the following assumptions: (i) A given gauge theory possesses simultaneously, at least, a non-trivial infrared (IR) fixed point and an ultraviolet (UV) one; (ii) Upon changing an external parameter of the theory, e.g. the number of flavors, at a critical value of this parameter the IR fixed point merges with the UV fixed point; (iii) This merging is sufficiently smooth that the nearby conformal phase is felt, in the conformally broken phase, for values of the external parameter near the phase transition.

Without loss of generality it is possible to model the beta function near the critical number of flavors as follows:

$$\beta_{MY} = -\alpha^2 \left(\alpha - 1 - \sqrt{\delta}\right)\left(\alpha - 1 + \sqrt{\delta}\right) = -\alpha^2((\alpha - 1)^2 - \delta). \qquad (2.41)$$

The double zero at the origin embodies asymptotic freedom and $\delta = n_f - n_f^c$. For positive values of δ the beta function possesses a non-trivial IR and UV fixed point at the following values of the coupling:

$$\alpha_{IR} = 1 - \sqrt{\delta}, \quad \text{and} \quad \alpha_{UV} = 1 + \sqrt{\delta}. \qquad (2.42)$$

At $n_f = n_f^c$ the fixed points merge and for $n_f < n_f^c$ the beta function looses the non-trivial fixed points. For negative δ within the following range:

$$-\frac{1}{8} < \delta \leq 0 \qquad (2.43)$$

there is a global maximum of the beta function at the origin, a local minimum at $\alpha = \frac{1}{4}(3 - \sqrt{1 + 8\delta})$ and a local maximum at $\alpha = \frac{1}{4}(3 + \sqrt{1 + 8\delta})$. For illustration we plot the beta function for different values of δ in Fig. 2.12. It is possible to find an analytical solution to the RG equation:

$$d \ln \mu = \frac{d\alpha}{\beta_{MY}}, \qquad (2.44)$$

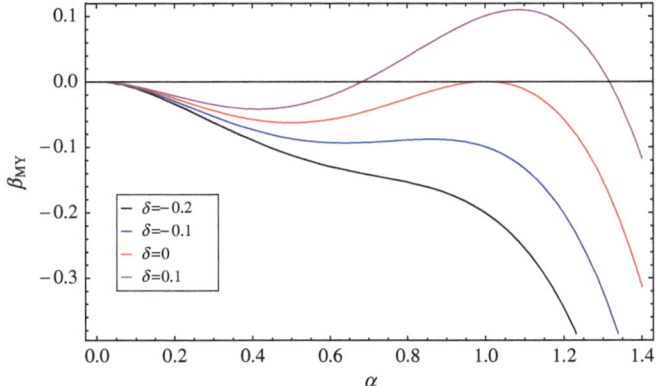

Fig. 2.12 β_{MY} for different values of $\delta = n_f - n_f^c$

reading:

$$\ln\frac{\mu}{\mu_0} = \frac{\alpha(1+\delta)\text{ArcTanh}\left[\frac{-1+\alpha}{\sqrt{\delta}}\right] + \sqrt{\delta}\left(1-\delta+\alpha\ln\left[\frac{(\alpha-1)^2-\delta}{\alpha^2}\right]\right)}{\alpha(-1+\delta)^2\sqrt{\delta}}\Bigg|_{\alpha(\mu_0)}^{\alpha(\mu)}.$$

(2.45)

If we were to consider the case of δ positive, but smaller than unity so that asymptotic freedom is kept, we would discover that there are three distinct branches. The one to the left of the IR fixed point, the one where α is in between the nontrivial IR and UV fixed point, and the one to the right of the nontrivial UV fixed point. To the left of the IR fixed point one starts the flow from any μ_0 sufficiently close to the trivial UV fixed point and one ends up at the attractive IR fixed point. Another *asymptotically safe* theory is defined in between the two non-trivial fixed points. In this region the coupling runs at low energies to the IR fixed point and raises at high energies till the non-trivial UV fixed point is reached. Finally, to the right of the non-trivial UV fixed point the theory, and hence beta function, runs in the deep infrared to increasingly large values of the coupling.

We turn now our attention to negative values of δ corresponding to the phase in which the beta function features no non-trivial fixed points. To elucidate the near-conformal dynamics we investigate the region $-1/8 < \delta < 0$. In particular we will consider the limit $-\delta = n_f^c - n_f \to 0$. In the deep infrared the coupling constant runs to infinity and we start the running in the UV near $\alpha = 1$ at μ_0. With these boundary conditions we find:

$$\Lambda_{MY} = \frac{\mu_0}{n_f^c - n_f}\exp\left[-\frac{\pi}{2\sqrt{n_f^c - n_f}}\right], \quad n_f \to n_f^c, n_f \leq n_f^c.$$

(2.46)

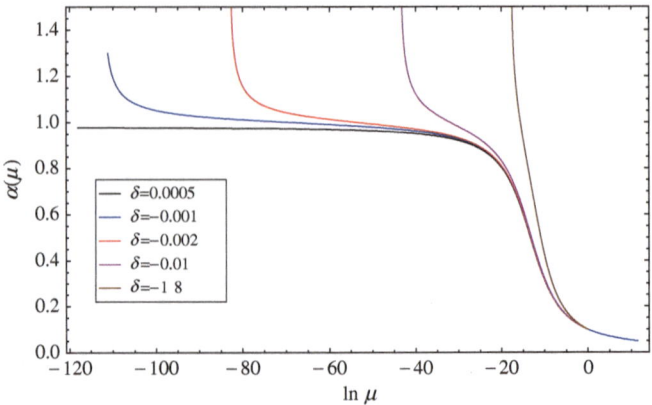

Fig. 2.13 Running of the coupling constant coming from β_{MY} for different values of $\delta = n_f - n_f^c$ within the *walking* regime. All the solutions are normalized at μ_0 such that $\alpha(\mu_0) = 0.1$ and μ is normalized to μ_0

Here Λ_{MY} is the infrared scale to be identified, for example, with a physical scale of the theory such as the mass of a hadron. This scale vanishes exponentially fast when approaching the critical number of flavors above which the infrared fixed point is generated. This exponential behavior is the essence of the Miransky scaling.

In Fig. 2.13 we plot the running of the coupling constant for different negative values of δ with the normalization condition $\alpha(\mu_0) = 0.1$. The figure visualizes the idea of walking dynamics introduced by Holdom [75, 76] and further crystallized in [63, 77]. In lay terms the coupling constant runs slowly, i.e. *walks*, towards the infrared value remaining near constant over a range of energies becoming wider and wider as one approaches, as function of δ, the double fixed point. Further assuming that the beta function corresponds to an underlying gauge theory featuring fermions we now determine the scaling behavior of the chiral condensate of the theory in the walking regime. Defining with γ the anomalous dimension of the mass of the Dirac fermion Q in a given representation of an underlying gauge group we have the following well-known RG equation:

$$\langle \bar{Q}Q \rangle_\mu = \exp\left(\int_\Lambda^\mu d(\ln \mu) \gamma(\alpha(\mu)) \right) \langle \bar{Q}Q \rangle_\Lambda$$

$$= \exp\left(\int_{\alpha(\Lambda)}^{\alpha(\mu)} d\alpha \frac{\gamma(\alpha)}{\beta(\alpha)} \right) \langle \bar{Q}Q \rangle_\Lambda \qquad (2.47)$$

relating the condensate at two different energies. Using β_{MY} case in the walking region we have

$$\langle \bar{Q}Q \rangle_\mu = \exp\left(\int_{\alpha(\Lambda)}^{\alpha(\mu)} d\alpha \frac{\gamma(\alpha)}{-\alpha^2((\alpha-1)^2 + |\delta|)} \right) \langle \bar{Q}Q \rangle_\Lambda$$

$$\simeq \exp\left(\gamma(1) \int_{\alpha(\Lambda)}^{\alpha(\mu)} d\alpha \frac{1}{\beta_{MY}} \right) \langle \bar{Q}Q \rangle_\Lambda = \left(\frac{\mu}{\Lambda} \right)^{\gamma(1)} \langle \bar{Q}Q \rangle_\Lambda . \qquad (2.48)$$

In the last passage we have used the definition of the beta function and, in the first step, assumed that the anomalous dimension of the mass operator is smooth across the phase transition. $\gamma(1) = \gamma(\alpha_{IR} = \alpha_{UV})$ is the value of the anomalous dimension at the merger. We have re-derived the power-law enhancement of the chiral condensate with the energy distinctive of walking dynamics. Since γ is evaluated at the fixed point its value is scheme-independent [6]. If we further model $\gamma = \alpha$ we have $\gamma(\alpha_{IR/UV}) = 1 \mp \sqrt{\delta}$, with $\gamma(\alpha_{IR}) + \gamma(\alpha_{UV}) = 2$.

2.6.2 Jumping Dynamics

The previous section embodies the standard paradigm of walking dynamics. However this picture is far from established analytically or via first principle lattice simulations in four dimensions, while lower dimensional examples exist [78]. It is therefore relevant to consider other theoretical scenarios and their impact on particle physics phenomenology. We start by observing that there is the logical possibility that the full beta function of the theory develops, at least, a zero in the denominator. This occurs exactly for supersymmetric gauge theories [67] and the all-orders beta function conjectured to be valid also for non-supersymmetric gauge theories with fermionic matter [4, 6]. Moreover it is reasonable to expect that the full perturbative and non-perturbative contributions to the beta function conspire to generate a non-trivial pole structure [79]. Whatever the pole structure is, if the underlying theory displays conformality, there will be also zeros in the numerator of the beta function associated to the non-trivial fixed point structure of the theory. Here we consider the simplest example in which the beta function has a simple nontrivial zero in the numerator and a simple pole. We will always assume the existence of the trivial double zero at the origin so that the beta function contains information about the asymptotically free nature of the theory. Without loss of generality we write:

$$\beta_{Jump} = -\alpha^2 \frac{1 - \delta - \alpha}{1 - \alpha} . \qquad (2.49)$$

By construction this beta function has a zero in the numerator for any δ which is to the left of the pole value $\alpha_{pole} = 1$ for δ positive and to the right for δ negative. Here we take again $\delta = n_f - n_f^c$. It is straightforward to show that this zero corresponds

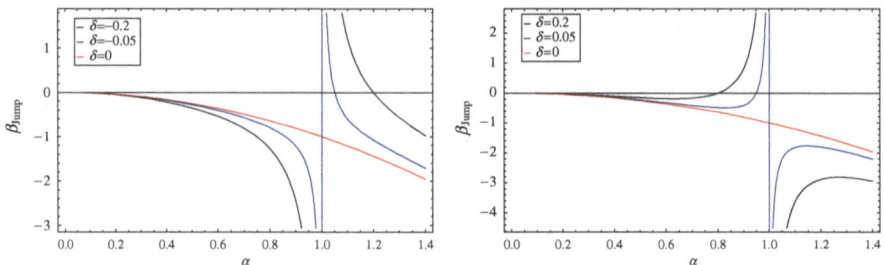

Fig. 2.14 β_{Jump} for different values of $\delta = n_f - n_f^c$

to an IR (UV) fixed point for $\delta > 0$ ($\delta < 0$):

$$\alpha_{IR(UV)} = 1 - \delta, \quad \delta > 0 \quad (\delta < 0) \quad \text{and} \quad |\delta| \leq 1. \tag{2.50}$$

Because of the presence of the pole the beta function describes two disconnected theories. One which is continuously connected to the asymptotically free underlying gauge theory and the other which is not. We plot the beta function for positive and negative values of delta in Fig. 2.14. At exactly $\delta = 0$ the numerator and denominator of the beta function cancel and we are left with $\beta_{Jump}^{\delta=0} = -\alpha^2$ which is the red curve in Fig. 2.14. What happens at the phase boundary? We will demonstrate that there is a sudden jump as we drop the number of flavors below the critical number of flavors (i.e. at $\delta = 0$) of the intrinsic physical scale of the theory.

We start by constructing the analytical solution for the RG equation of the coupling which reads:

$$\ln \frac{\mu}{\mu_0} = \frac{1}{\alpha(1-\delta)}\Big|_{\alpha(\mu_0)}^{\alpha(\mu)} + \frac{\delta}{(1-\delta)^2} \ln \left[\frac{1-\delta-\alpha}{\alpha}\right]\Big|_{\alpha(\mu_0)}^{\alpha(\mu)}. \tag{2.51}$$

Holding fixed, as done for the Miransky scaling case, the coupling constant at a given renormalization scale one observes that the newly generated scale increases with decreasing the number of flavors below the conformal window in the following way:

$$\Lambda_{Jump} = \Lambda_c \left[1 - (n_f^c - n_f) \ln \left(n_f^c - n_f\right)\right], n_f \to n_f^c, n_f \leq n_f^c. \tag{2.52}$$

$\Lambda_c = \mu_0 \exp\left[\frac{\ln \alpha_0}{\alpha_0}\right]$ is the renormalization group invariant scale of the theory at the critical number of flavors. However for $n_f > n_f^c$ no infrared scale is generated and necessarily there must be a *jump* in the spectrum from Λ_c to zero (Fig. 2.15). This result shows that β_{MY} and β_{Jump} describe two distinct physical systems. For illustration we summarize in Fig. 2.16 the behavior of the physical scale of the theory, as function of number of flavors, for Miransky scaling and jumping dynamics. To compare the two scaling laws we normalized the two scales at a given value of n_f.

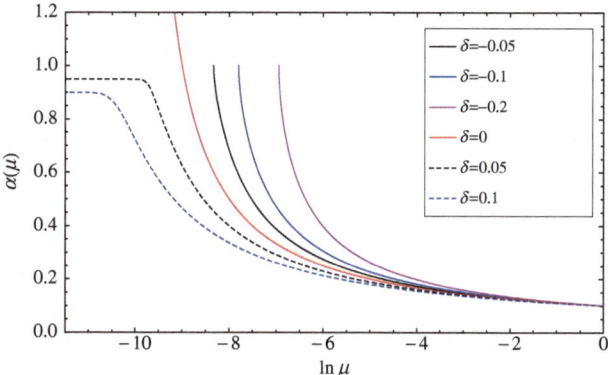

Fig. 2.15 Running of the coupling constant due to β_{Jump} for different values of $\delta = n_f - n_f^c$. All the solutions are normalized at μ_0 such that $\alpha(\mu_0) = 0.1$, and of course μ is normalized to μ_0

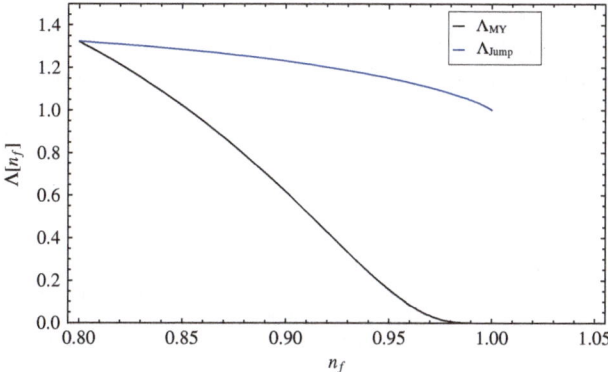

Fig. 2.16 Mass gap dependence on the number of flavors for the Miransky scaling (*lower black curve*) and for the jumping dynamics (*upper blue curve*). We have taken for illustration $n_f^c = 1$

Can jumping dynamics be used for models of dynamical electroweak symmetry breaking? At the conformal boundary the dynamics is QCD like and therefore one observes only a logarithmic enhancement of the condensate of the type $\langle \bar{Q}Q \rangle_\mu \simeq \gamma(1) \ \ln\left(\frac{\mu}{\Lambda}\right) \langle \bar{Q}Q \rangle_\Lambda$. The jumping dynamics does not lead to power-law enhancement of the chiral condensate required for *walking* technicolor. Furthermore the S-parameter in the jumping scenario automatically respects the lower bound put forward in [80] given that, opportunely normalized, at the phase boundary is as small as the one for QCD. Henceforth the answer to the original question is that one can break the electroweak theory via jumping dynamics but cannot accommodate the generation of the standard model fermion masses following the walking paradigm nor drastically reduce the QCD-like S-parameter.

We have shown that it is possible to devise a simple framework according to which the approach to the long distance conformality does not display any sign of walking dynamics. All the current lattice investigations of the conformal window are not able to differentiate walking from jumping. The reasons being that: (i) The lattice results, for the moment, are performed for a fixed number of flavors and therefore either there is a nonzero infrared scale or the theory is conformal; (ii) The precise determination of the chiral condensate is not a simple task making harder to disentangle a power-law from a logarithmic enhancement of the condensate as function of the renormalization scale as well as the number of flavors; (iii) Even if the underlying dynamics is of walking type (with or without the introduction of four-fermion interactions) the extension of the region in the number of flavors and four-fermion coupling is not known and might be tiny; (iv) Measuring a large anomalous dimension of the mass is encouraging but alone insufficient to demonstrate the existence of walking dynamics.

Because of the discontinuity of the order parameter at the conformal phase transition, i.e. of the vacuum expectation value of the trace of the improved energy momentum tensor which is proportional to the intrinsic scale of the theory, jumping dynamics corresponds to a first order conformal phase transition. First order phase transitions are common in nature and therefore we expect jumping dynamics to constitute a likely scenario with inevitable important consequences on a large number of research fields ranging from a better understanding of strong dynamics and its holographic engineering to the construction of sensible extensions of the standard model.

References

1. F. Sannino, K. Tuominen, Phys. Rev. D **71**, 051901 (2005) [arXiv:hep-ph/0405209]
2. D.D. Dietrich, F. Sannino, Phys. Rev. D **75**, 085018 (2007) [arXiv:hep-ph/0611341]
3. T.A. Ryttov, F. Sannino, Phys. Rev. D **76**, 105004 (2007) [arXiv:0707.3166 [hep-th]]
4. T.A. Ryttov, F. Sannino, Phys. Rev. D **78**, 065001 (2008) [arXiv:0711.3745 [hep-th]]
5. F. Sannino, arXiv:0804.0182 [hep-ph]
6. C. Pica, F. Sannino, Phys. Rev. D **83**, 116001 (2011) [arXiv:1011.3832 [hep-ph]]
7. F. Sannino, Phys. Rev. D **72**, 125006 (2005) [hep-th/0507251]
8. F. Sannino, Phys. Rev. D **79**, 096007 (2009) [arXiv:0902.3494 [hep-ph]]
9. T.A. Ryttov, R. Shrock, Phys. Rev. D **81**, 116003 (2010) [Erratum-ibid. D **82**, 059903 (2010)] [arXiv:1006.0421 [hep-ph]]
10. N. Chen, T.A. Ryttov, R. Shrock, Phys. Rev. D **82**, 116006 (2010) [arXiv:1010.3736 [hep-ph]]
11. M. Mojaza, C. Pica, F. Sannino, Phys. Rev. D **82**, 116009 (2010) [arXiv:1010.4798 [hep-ph]]
12. O. Antipin, M. Mojaza, F. Sannino, Phys. Lett. B **712**, 119 (2012) [arXiv:1107.2932 [hep-ph]]
13. O. Antipin, S. Di Chiara, M. Mojaza, E. Molgaard, F. Sannino, arXiv:1205.6157 [hep-ph]
14. M. Mojaza, C. Pica, T.A. Ryttov, F. Sannino, arXiv:1206.2652 [hep-ph]
15. C. Pica, F. Sannino, Phys. Rev. D **83**, 035013 (2011) [arXiv:1011.5917 [hep-ph]]
16. T.A. Ryttov, R. Shrock, Phys. Rev. D **83**, 056011 (2011) [arXiv:1011.4542 [hep-ph]]
17. T. Appelquist, K.D. Lane, U. Mahanta, Phys. Rev. Lett. **61**, 1553 (1988)
18. A.G. Cohen, H. Georgi, Nucl. Phys. B **314**, 7 (1989)
19. V.A. Miransky, K. Yamawaki, Phys. Rev. D **55**, 5051 (1997) [Erratum-ibid. D **56**, 3768 (1997)] [arXiv:hep-th/9611142]
20. K.A. Intriligator, N. Seiberg, Nucl. Phys. Proc. Suppl. **45BC**, 1 (1996) [arXiv:hep-th/9509066]
21. S. Weinberg, Ultraviolet divergences in quantum theories of gravitation, in *General Relativity: An Einstein Centenary Survey*, ed. by S.W. Hawking, W. Israel (Cambridge University Press, Cambridge, 1979), p. 790

22. D.F. Litim, Phil. Trans. Roy. Soc. Lond. A **369**, 2759 (2011) [arXiv:1102.4624 [hep-th]]
23. T. van Ritbergen, J.A.M. Vermaseren, S.A. Larin, Phys. Lett. B **400**, 379 (1997) [hep-ph/9701390]
24. J.A.M. Vermaseren, S.A. Larin, T. van Ritbergen, Phys. Lett. B **405**, 327 (1997) [hep-ph/9703284]
25. L. Li, Y. Meurice, Phys. Rev. D **71**, 016008 (2005) [hep-lat/0410029]
26. T. Banks, A. Zaks, Nucl. Phys. B **196**, 189 (1982)
27. F. Sannino, Phys. Rev. D **80**, 065011 (2009) [arXiv:0907.1364 [hep-th]]
28. F. Sannino, arXiv:0909.4584 [hep-th]
29. D.K. Hong, S.D.H. Hsu, F. Sannino, Phys. Lett. B **597**, 89 (2004) [arXiv:hep-ph/0406200]
30. D.D. Dietrich, F. Sannino, K. Tuominen, Phys. Rev. D **72**, 055001 (2005) [arXiv:hep-ph/0505059]
31. S. Catterall, F. Sannino, Phys. Rev. D **76**, 034504 (2007) [arXiv:0705.1664 [hep-lat]]
32. L. Del Debbio, M.T. Frandsen, H. Panagopoulos, F. Sannino, JHEP **0806**, 007 (2008) [arXiv:0802.0891 [hep-lat]]
33. Y. Shamir, B. Svetitsky, T. DeGrand, Phys. Rev. D **78**, 031502 (2008) [arXiv:0803.1707 [hep-lat]]
34. A. Deuzeman, M.P. Lombardo, E. Pallante, Phys. Lett. B **670**, 41 (2008) [arXiv:0804.2905 [hep-lat]]
35. L. Del Debbio, A. Patella, C. Pica, Phys. Rev. D **81**, 094503 (2010) [arXiv:0805.2058 [hep-lat]]
36. S. Catterall, J. Giedt, F. Sannino, J. Schneible, JHEP **0811**, 009 (2008) [arXiv:0807.0792 [hep-lat]]
37. L. Del Debbio, A. Patella, C. Pica, PoS **LATTICE2008**, 064 (2008) [arXiv:0812.0570 [hep-lat]]
38. T. DeGrand, Y. Shamir, B. Svetitsky, Phys. Rev. D **79**, 034501 (2009) [arXiv:0812.1427 [hep-lat]]
39. A.J. Hietanen, J. Rantaharju, K. Rummukainen, K. Tuominen, JHEP **0905**, 025 (2009) [arXiv:0812.1467 [hep-lat]]
40. T. Appelquist, G.T. Fleming, E.T. Neil, Phys. Rev. D **79**, 076010 (2009) [arXiv:0901.3766 [hep-ph]]
41. A.J. Hietanen, K. Rummukainen, K. Tuominen, Phys. Rev. D **80**, 094504 (2009) [arXiv:0904.0864 [hep-lat]]
42. A. Deuzeman, M.P. Lombardo, E. Pallante, Phys. Rev. D **82**, 074503 (2010) [arXiv:0904.4662 [hep-ph]]
43. A. Hasenfratz, Phys. Rev. D **80**, 034505 (2009) [arXiv:0907.0919 [hep-lat]]
44. L. Del Debbio, B. Lucini, A. Patella, C. Pica, A. Rago, Phys. Rev. D **80**, 074507 (2009) [arXiv:0907.3896 [hep-lat]]
45. Z. Fodor, K. Holland, J. Kuti, D. Nogradi, C. Schroeder, Phys. Lett. B **681**, 353 (2009) [arXiv:0907.4562 [hep-lat]]
46. Z. Fodor, K. Holland, J. Kuti, D. Nogradi, C. Schroeder, JHEP **0911**, 103 (2009) [arXiv:0908.2466 [hep-lat]]
47. T. DeGrand, Phys. Rev. D **80**, 114507 (2009) [arXiv:0910.3072 [hep-lat]]
48. S. Catterall, J. Giedt, F. Sannino, J. Schneible, arXiv:0910.4387 [hep-lat]
49. F. Bursa, L. Del Debbio, L. Keegan, C. Pica, T. Pickup, Phys. Rev. D **81**, 014505 (2010) [arXiv:0910.4535 [hep-ph]]
50. E. Bilgici et al., Phys. Rev. D **80**, 034507 (2009) [arXiv:0902.3768 [hep-lat]]
51. J.B. Kogut, D.K. Sinclair, Phys. Rev. D **81**, 114507 (2010) [arXiv:1002.2988 [hep-lat]]
52. A. Hasenfratz, Phys. Rev. D **82**, 014506 (2010) [arXiv:1004.1004 [hep-lat]]
53. L. Del Debbio, B. Lucini, A. Patella, C. Pica, A. Rago, Phys. Rev. D **82**, 014509 (2010) [arXiv:1004.3197 [hep-lat]]
54. L. Del Debbio, B. Lucini, A. Patella, C. Pica, A. Rago, Phys. Rev. D **82**, 014510 (2010) [arXiv:1004.3206 [hep-lat]]

55. S. Catterall, L. Del Debbio, J. Giedt, L. Keegan, PoS **LATTICE2010**, 057 (2010) [arXiv:1010.5909 [hep-ph]]
56. Z. Fodor, K. Holland, J. Kuti, D. Nogradi, C. Schroeder, arXiv:1103.5998 [hep-lat]
57. R. Lewis, C. Pica, F. Sannino, Phys. Rev. D **85**, 014504 (2012) [arXiv:1109.3513 [hep-ph]]
58. Z. Fodor, K. Holland, J. Kuti, D. Nogradi, C. Schroeder, C.H. Wong, arXiv:1205.1878 [hep-lat]
59. J. Giedt, E. Weinberg, Phys. Rev. D **85**, 097503 (2012) [arXiv:1201.6262 [hep-lat]]
60. B. Holdom, Phys. Lett. B **694**, 74 (2010) [arXiv:1006.2119 [hep-ph]]
61. T. Maskawa, H. Nakajima, Prog. Theor. Phys. **52**, 1326 (1974)
62. R. Fukuda, T. Kugo, Nucl. Phys. B **117**, 250 (1976)
63. K. Yamawaki, M. Bando, K.i. Matumoto, Phys. Rev. Lett. **56**, 1335 (1986)
64. N. Seiberg, Phys. Rev. D **49**, 6857 (1994) [arXiv:hep-th/9402044]
65. N. Seiberg, Nucl. Phys. B **435**, 129 (1995) [arXiv:hep-th/9411149]
66. G. 't Hooft, C. Itzykson, A. Jaffe, H. Lehmann, P.K. Mitter, I.M. Singer, R. Stora (eds.), *Recent Developments in Gauge Theories*. Nato Advanced Study Institutes Series: Series B, Physics, vol. 59 (Plenum, New York, 1980), p. 438
67. V.A. Novikov, M.A. Shifman, A.I. Vainshtein, V.I. Zakharov, Nucl. Phys. B **229**, 381 (1983)
68. M.A. Shifman, A.I. Vainshtein, Nucl. Phys. B **277**, 456 (1986) [Sov. Phys. JETP **64**, 428 (1986 ZETFA,91,723–744.1986)]
69. D.R.T. Jones, Phys. Lett. B **123**, 45 (1983)
70. J. Terning, Phys. Rev. Lett. **80**, 2517 (1998) [arXiv:hep-th/9706074]
71. F. Sannino, Acta Phys. Polon. B **40**, 3533 (2009) [arXiv:0911.0931 [hep-ph]]
72. F. Sannino, Phys. Rev. Lett. **105**, 232002 (2010) [arXiv:1007.0254 [hep-ph]]
73. V.A. Miransky, K. Yamawaki, Mod. Phys. Lett. A **4**, 129 (1989)
74. F. Sannino, arXiv:1205.4246 [hep-ph]
75. B. Holdom, Phys. Rev. D **24**, 1441 (1981)
76. B. Holdom, Phys. Lett. B **150**, 301 (1985)
77. T.W. Appelquist, D. Karabali, L.C.R. Wijewardhana, Phys. Rev. Lett. **57**, 957 (1986)
78. P. de Forcrand, M. Pepe, U.-J. Wiese, arXiv:1204.4913 [hep-lat]
79. F.A. Chishtie, V. Elias, V.A. Miransky, T.G. Steele, Prog. Theor. Phys. **104**, 603 (2000) [hep-ph/9905291]
80. F. Sannino, Phys. Rev. D **82**, 081701 (2010) [arXiv:1006.0207 [hep-lat]]

Chapter 3
Minimal Technicolor Models: Toccata and Fugue

Abstract In the prelude we introduced the basic concepts about technicolor and argued for the need of new strongly coupled dynamics. In the previous interlude section we argued for the existence of such a new type of dynamics. Here we construct a few explicit model examples. The associated LHC phenomenology is reported in the *Discovering Technicolor* report [1].

3.1 Minimal Technicolor

We start by making the observation that the simplest technicolor models has N_{Tf} Dirac fermions in the fundamental representation of $SU(N)$. These models, when extended to accommodate the fermion masses through the extended technicolor interactions, suffer from large flavor changing neutral currents. This problem is alleviated if the number of flavors is sufficiently large such that the theory is almost conformal. This is estimated to happen for $N_{Tf} \sim 4N$ [2] as also summarized in the section dedicated to the Phase Diagram of strongly interacting theories. This, in turn, implies a large contribution to the oblique parameter S (within naive estimates) [3]. Although near the conformal window [4, 5] the S parameter is reduced due to non-perturbative corrections, it is still too large if the model has a large particle content. In addition, such models may have a large number of pseudo Nambu-Goldstone bosons. By choosing a higher dimensional technicolor representation for the new technifermions one can overcome these problems [3, 6].

To have a very low S parameter one would ideally have a technicolor theory which with only one doublet breaks dynamically the electroweak theory but at the same time being walking to reduce the S parameter. The walking nature then also enhances the scale responsible for the fermion mass generation.

According to the phase diagram exhibited earlier the promising candidate theories with the properties required are either theories with fermions in the adjoint representation or two index symmetric one. In Table 3.1 we present the generic S-type theory.

F. Sannino, *Dynamical Stabilization of the Fermi Scale*, SpringerBriefs in Physics, 61
DOI: 10.1007/978-3-642-33341-5_3, © The Author(s) 2013

Table 3.1 Schematic representation of a generic nonsupersymmetric vector like $SU(N)$ gauge theory with matter content in the two-index representation

	$SU(N)$	$SU_L(N_{Tf})$	$SU_R(N_{Tf})$	$U_V(1)$	$U_A(1)$
Q_L	☐☐	☐	1	1	1
$\tilde{Q}_R$	☐☐	1	☐	-1	1
G_μ	Adj	0	0	0	0

Here $Q_{L(R)}$ are Weyl fermions

The relevant feature, found first in [6] using the ladder approximation, is that the S-type theories can be near conformal already at $N_{Tf} = 2$ when $N = 2$ or 3. This should be contrasted with theories in which the fermions are in the fundamental representation for which the minimum number of flavors required to reach the conformal window is eight for $N = 2$. This last statement is supported by the all order beta function results [7] as well as lattice simulations [8–11]. The critical value of flavors increases with the number of colors for the gauge theory with S-type matter: the limiting value is 4.15 at large N.

The situation is different for the theory with A-type matter. Here the critical number of flavors increases when decreasing the number of colors. The maximum value of about $N_{Tf} = 12$ is obtained—in the ladder approximation—for $N = 3$, i.e. standard QCD. In Ref. [3] it has been argued that the nearly conformal A-type theories have, already at the perturbative level, a very large S parameter with respect to the experimental data. These theories can be re-considered if one gauges under the electroweak symmetry only a part of the flavor symmetries as we shall see in the section dedicated to *partially gauged* technicolor.

3.2 Minimal Walking Technicolor (MWT)

The dynamical sector we consider, which underlies the Higgs mechanism, is an SU(2) technicolor gauge theory with two adjoint technifermions [6]. The theory is asymptotically free if the number of flavors n_f is less than 2.75 according to the ladder approximation. Lattice results support the conformal or near conformal behavior of this theory. In any event the symmetries and properties of this model make it ideal for a comprehensive study for LHC physics. The all order beta function prediction is that this gauge theory is, in fact, conformal. In this case we can couple another non-conformal sector to this gauge theory and push it away from the fixed point.

The two adjoint fermions are conveniently written as

$$Q_L^a = \begin{pmatrix} U^a \\ D^a \end{pmatrix}_L, \quad U_R^a, \quad D_R^a, \quad a = 1, 2, 3, \tag{3.1}$$

with a being the adjoint color index of SU(2). The left handed fields are arranged in three doublets of the $SU(2)_L$ weak interactions in the standard fashion. The condensate is $\langle \bar{U}U + \bar{D}D \rangle$ which correctly breaks the electroweak symmetry as already argued for ordinary QCD.

The model as described so far suffers from the Witten topological anomaly [12]. However, this can easily be solved by adding a new weakly charged fermionic doublet which is a technicolor singlet [13]. Schematically:

$$L_L = \begin{pmatrix} N \\ E \end{pmatrix}_L , \quad N_R, E_R. \tag{3.2}$$

In general, the gauge anomalies cancel using the following generic hypercharge assignment

$$Y(Q_L) = \frac{y}{2}, \quad Y(U_R, D_R) = \left(\frac{y+1}{2}, \frac{y-1}{2} \right), \tag{3.3}$$

$$Y(L_L) = -3\frac{y}{2}, \quad Y(N_R, E_R) = \left(\frac{-3y+1}{2}, \frac{-3y-1}{2} \right), \tag{3.4}$$

where the parameter y can take any real value [13]. In our notation the electric charge is $Q = T_3 + Y$, where T_3 is the weak isospin generator. One recovers the SM hypercharge assignment for $y = 1/3$ (Fig. 3.1).

To discuss the symmetry properties of the theory it is convenient to use the Weyl basis for the fermions and arrange them in the following vector transforming according to the fundamental representation of SU(4)

$$Q = \begin{pmatrix} U_L \\ D_L \\ -i\sigma^2 U_R^* \\ -i\sigma^2 D_R^* \end{pmatrix}, \tag{3.5}$$

where U_L and D_L are the left handed techniup and technidown, respectively and U_R and D_R are the corresponding right handed particles. Assuming the standard breaking to the maximal diagonal subgroup, the SU(4) symmetry spontaneously breaks to $SO(4)$. Such a breaking is driven by the following condensate

$$\langle Q_i^\alpha Q_j^\beta \varepsilon_{\alpha\beta} E^{ij} \rangle = -2\langle \bar{U}_R U_L + \bar{D}_R D_L \rangle , \tag{3.6}$$

where the indices $i, j = 1, \ldots, 4$ denote the components of the tetraplet of Q, and the Greek indices indicate the ordinary spin. The matrix E is a 4×4 matrix defined in terms of the 2-dimensional unit matrix as

$$E = \begin{pmatrix} 0 & \mathbb{1} \\ \mathbb{1} & 0 \end{pmatrix}. \tag{3.7}$$

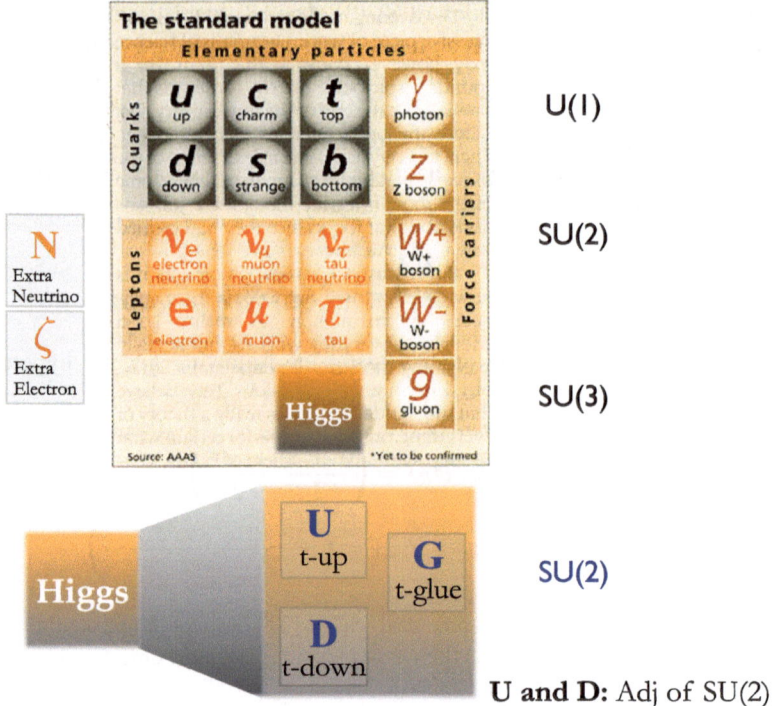

Fig. 3.1 Cartoon of the minimal walking technicolor model extension of the SM

Here $\varepsilon_{\alpha\beta} = -i\sigma^2_{\alpha\beta}$ and $\langle U^\alpha_L U_R^{*\beta}\varepsilon_{\alpha\beta}\rangle = -\langle \overline{U}_R U_L\rangle$. A similar expression holds for the D techniquark. The above condensate is invariant under an $SO(4)$ symmetry. This leaves us with nine broken generators with associated Goldstone bosons.

Replacing the Higgs sector of the SM with the MWT the Lagrangian now reads:

$$\mathscr{L}_H \rightarrow -\frac{1}{4}\mathscr{F}^a_{\mu\nu}\mathscr{F}^{a\mu\nu} + i\bar{Q}_L\gamma^\mu D_\mu Q_L + i\bar{U}_R\gamma^\mu D_\mu U_R + i\bar{D}_R\gamma^\mu D_\mu D_R$$
$$+i\bar{L}_L\gamma^\mu D_\mu L_L + i\bar{N}_R\gamma^\mu D_\mu N_R + i\bar{E}_R\gamma^\mu D_\mu E_R \qquad (3.8)$$

with the technicolor field strength $\mathscr{F}^a_{\mu\nu} = \partial_\mu\mathscr{A}^a_\nu - \partial_\nu\mathscr{A}^a_\mu + g_{TC}\varepsilon^{abc}\mathscr{A}^b_\mu\mathscr{A}^c_\nu$, $a, b, c = 1, \ldots, 3$. For the left handed techniquarks the covariant derivative is:

$$D_\mu Q^a_L = \left(\delta^{ac}\partial_\mu + g_{TC}\mathscr{A}^b_\mu\varepsilon^{abc} - i\frac{g}{2}\mathbf{W}_\mu \cdot \tau\delta^{ac} - ig'\frac{y}{2}B_\mu\delta^{ac}\right)Q^c_L. \qquad (3.9)$$

$\mathscr{A}_\mu$ are the techni gauge bosons, W_μ are the gauge bosons associated to SU(2)$_L$ and B_μ is the gauge boson associated to the hypercharge. τ^a are the Pauli matrices and ε^{abc} is the fully antisymmetric symbol. In the case of right handed techniquarks the third term containing the weak interactions disappears and the hypercharge $y/2$ has to be replaced according to whether it is an up or down techniquark. For the left-handed leptons the second term containing the technicolor interactions disappears and $y/2$ changes to $-3y/2$. Only the last term is present for the right handed leptons with an appropriate hypercharge assignment.

3.3 Low Energy Theory for MWT

We construct the effective theory for MWT including composite scalars and vector bosons, their self interactions, and their interactions with the electroweak gauge fields and the SM fermions.

3.3.1 Scalar Sector

The relevant effective theory for the Higgs sector at the electroweak scale consists, in our model, of a composite Higgs σ and its pseudoscalar partner Θ, as well as nine pseudoscalar Goldstone bosons and their scalar partners. The recent discovery of a Higgs-like state would naturally fit within this framework.[1] These states can be assembled in the matrix

$$M = \left[\frac{\sigma + i\Theta}{2} + \sqrt{2}(i\Pi^a + \tilde{\Pi}^a)X^a \right] E , \qquad (3.10)$$

which transforms under the full $SU(4)$ group according to

$$M \rightarrow uMu^T, \quad \text{with} \quad u \in SU(4). \qquad (3.11)$$

The X^a's, $a = 1, \ldots, 9$ are the generators of the $SU(4)$ group which do not leave the Vacuum Expectation Value (VEV) of M invariant

$$\langle M \rangle = \frac{v}{2} E . \qquad (3.12)$$

Note that the notation used is such that σ is a *scalar* while the Π^a's are *pseudoscalars*. It is convenient to separate the fifteen generators of $SU(4)$ into the six that leave the vacuum invariant, S^a, and the remaining nine that do not, X^a. Then the S^a generators of the $SO(4)$ subgroup satisfy the relation

$$S^a E + E S^{aT} = 0, \quad \text{with} \quad a = 1, \ldots, 6, \qquad (3.13)$$

---

[1] In fact the original name for this model was *Light Composite Higgs* where the lightness of the composite Higgs was argued on the near-conformal nature of the model [13].

so that $uEu^T = E$, for $u \in SO(4)$. The explicit realization of the generators and the embedding of the electroweak generators in the $SU(4)$ algebra are shown in Appendix A.3. With the tilde fields included, the matrix M is invariant in form under $U(4) \equiv SU(4) \times U(1)_A$, rather than just $SU(4)$. However the $U(1)_A$ axial symmetry is anomalous, and is therefore broken at the quantum level.

The connection between the composite scalars and the underlying techniquarks can be derived from the transformation properties under $SU(4)$, by observing that the elements of the matrix M transform like techniquark bilinears:

$$M_{ij} \sim Q_i^\alpha Q_j^\beta \varepsilon_{\alpha\beta} \quad \text{with} \quad i, j = 1 \ldots 4. \tag{3.14}$$

Using this expression, and the basis matrices given in Appendix A.3, the scalar fields can be related to the wavefunctions of the techniquark bound states. This gives the following charge eigenstates:

$$v + H \equiv \sigma \sim \overline{U}U + \overline{D}D, \qquad \Theta \sim i\left(\overline{U}\gamma^5 U + \overline{D}\gamma^5 D\right),$$

$$A^0 \equiv \tilde{\Pi}^3 \sim \overline{U}U - \overline{D}D, \qquad \Pi^0 \equiv \Pi^3 \sim i\left(\overline{U}\gamma^5 U - \overline{D}\gamma^5 D\right),$$

$$A^+ \equiv \frac{\tilde{\Pi}^1 - i\tilde{\Pi}^2}{\sqrt{2}} \sim \overline{D}U, \quad \Pi^+ \equiv \frac{\Pi^1 - i\Pi^2}{\sqrt{2}} \sim i\overline{D}\gamma^5 U, \tag{3.15}$$

$$A^- \equiv \frac{\tilde{\Pi}^1 + i\tilde{\Pi}^2}{\sqrt{2}} \sim \overline{U}D, \quad \Pi^- \equiv \frac{\Pi^1 + i\Pi^2}{\sqrt{2}} \sim i\overline{U}\gamma^5 D,$$

for the technimesons, and

$$\Pi_{UU} \equiv \frac{\Pi^4 + i\Pi^5 + \Pi^6 + i\Pi^7}{2} \sim U^T C U,$$

$$\Pi_{DD} \equiv \frac{\Pi^4 + i\Pi^5 - \Pi^6 - i\Pi^7}{2} \sim D^T C D,$$

$$\Pi_{UD} \equiv \frac{\Pi^8 + i\Pi^9}{\sqrt{2}} \sim U^T C D,$$

$$\tilde{\Pi}_{UU} \equiv \frac{\tilde{\Pi}^4 + i\tilde{\Pi}^5 + \tilde{\Pi}^6 + i\tilde{\Pi}^7}{2} \sim iU^T C\gamma^5 U, \tag{3.16}$$

$$\tilde{\Pi}_{DD} \equiv \frac{\tilde{\Pi}^4 + i\tilde{\Pi}^5 - \tilde{\Pi}^6 - i\tilde{\Pi}^7}{2} \sim iD^T C\gamma^5 D,$$

$$\tilde{\Pi}_{UD} \equiv \frac{\tilde{\Pi}^8 + i\tilde{\Pi}^9}{\sqrt{2}} \sim iU^T C\gamma^5 D,$$

for the technibaryons, where $U \equiv (U_L, U_R)^T$ and $D \equiv (D_L, D_R)^T$ are Dirac technifermions, and C is the charge conjugation matrix, needed to form Lorentz-invariant objects. To these technibaryon charge eigenstates we must add the corresponding charge conjugate states (e.g. $\Pi_{UU} \rightarrow \Pi_{\overline{UU}}$).

Three of the nine Goldstone bosons ($\Pi^{\pm}$, Π^0) associated with the relative broken generators become the longitudinal degrees of freedom of the massive weak gauge bosons, while the extra six Goldstone bosons will acquire a mass due to ETC interactions as well as the electroweak interactions per se. Using a bottom up approach we will not commit to a specific ETC theory but limit ourself to introduce the minimal low energy operators needed to construct a phenomenologically viable theory. The new Higgs Lagrangian is

$$\mathscr{L}_{\text{Higgs}} = \frac{1}{2}\text{Tr}\left[D_\mu M D^\mu M^\dagger\right] - \mathscr{V}(M) + \mathscr{L}_{\text{ETC}}, \tag{3.17}$$

where the potential reads

$$\mathscr{V}(M) = -\frac{m_M^2}{2}\text{Tr}[MM^\dagger] + \frac{\lambda}{4}\text{Tr}\left[MM^\dagger\right]^2 + \lambda'\text{Tr}\left[MM^\dagger MM^\dagger\right]$$
$$- 2\lambda''\left[\text{Det}(M) + \text{Det}(M^\dagger)\right], \tag{3.18}$$

and $\mathscr{L}_{\text{ETC}}$ contains all terms which are generated by the ETC interactions, and not by the chiral symmetry breaking sector. Notice that the determinant terms (which are renormalizable) explicitly break the $U(1)_A$ symmetry, and give mass to Θ, which would otherwise be a massless Goldstone boson.

In order to give masses to the remaining uneaten Goldstone boson we add this term which is generated in the ETC sector:

$$\mathscr{L}_{\text{ETC}} \supset \frac{m_{\text{ETC}}^2}{4}\text{Tr}\left[MBM^\dagger B + MM^\dagger\right], \tag{3.19}$$

and $B \equiv 2\sqrt{2}S^4$ is a specific generator in the $SU(4)$ algebra.

The potential $\mathscr{V}(M)$ produces a VEV which parameterizes the techniquark condensate, and spontaneously breaks $SU(4)$ to $SO(4)$. In terms of the model parameters the VEV is

$$v^2 = \langle\sigma\rangle^2 = \frac{m_M^2}{\lambda + \lambda' - \lambda''}, \tag{3.20}$$

while the Higgs mass is

$$M_H^2 = 2\,m_M^2. \tag{3.21}$$

The linear combination $\lambda + \lambda' - \lambda''$ corresponds to the Higgs self coupling in the SM. The three pseudoscalar mesons $\Pi^{\pm}$, Π^0 correspond to the three massless Goldstone bosons which are absorbed by the longitudinal degrees of freedom of the $W^{\pm}$ and Z boson. The remaining six uneaten Goldstone bosons are technibaryons, and all acquire tree-level degenerate mass through the ETC interaction in (3.19):

$$M_{\Pi_{UU}}^2 = M_{\Pi_{UD}}^2 = M_{\Pi_{DD}}^2 = m_{\text{ETC}}^2. \tag{3.22}$$

The remaining scalar and pseudoscalar masses are

$$M_\Theta^2 = 4v^2\lambda''$$
$$M_{A^\pm}^2 = M_{A^0}^2 = 2v^2\left(\lambda' + \lambda''\right) \tag{3.23}$$

for the technimesons, and

$$M_{\tilde{\Pi}_{UU}}^2 = M_{\tilde{\Pi}_{UD}}^2 = M_{\tilde{\Pi}_{DD}}^2 = m_{\text{ETC}}^2 + 2v^2\left(\lambda' + \lambda''\right), \tag{3.24}$$

for the technibaryons. To gain insight on some of the mass relations one can use [3].

3.3.2 Vector Bosons

The composite vector bosons of a theory with a global $SU(4)$ symmetry are conveniently described by the four-dimensional traceless Hermitian matrix

$$A^\mu = A^{a\mu}\,T^a, \tag{3.25}$$

where T^a are the $SU(4)$ generators: $T^a = S^a$, for $a = 1, \ldots, 6$, and $T^{a+6} = X^a$, for $a = 1, \ldots, 9$. Under an arbitrary $SU(4)$ transformation, A^μ transforms like

$$A^\mu \rightarrow u\,A^m uu^\dagger, \quad \text{where } u \in SU(4). \tag{3.26}$$

Equation (3.26), together with the tracelessness of the matrix A_μ, gives the connection with the techniquark bilinears:

$$A_i^{\mu,j} \sim Q_i^\alpha \sigma_{\alpha\dot\beta}^\mu \bar{Q}^{\dot\beta,j} - \frac{1}{4}\delta_i^j\,Q_k^\alpha \sigma_{\alpha\dot\beta}^\mu \bar{Q}^{\dot\beta,k}. \tag{3.27}$$

Then we find the following relations between the charge eigenstates and the wavefunctions of the composite objects:

$$\begin{aligned}
v^{0\mu} &\equiv A^{3\mu} \sim \bar{U}\gamma^\mu U - \bar{D}\gamma^\mu D, & a^{0\mu} &\equiv A^{9\mu} \sim \bar{U}\gamma^\mu\gamma^5 U - \bar{D}\gamma^\mu\gamma^5 D \\
v^{+\mu} &\equiv \frac{A^{1\mu} - iA^{2\mu}}{\sqrt{2}} \sim \bar{D}\gamma^\mu U, & a^{+\mu} &\equiv \frac{A^{7\mu} - iA^{8\mu}}{\sqrt{2}} \sim \bar{D}\gamma^\mu\gamma^5 U \quad (3.28) \\
v^{-\mu} &\equiv \frac{A^{1\mu} + iA^{2\mu}}{\sqrt{2}} \sim \bar{U}\gamma^\mu D, & a^{-\mu} &\equiv \frac{A^{7\mu} + iA^{8\mu}}{\sqrt{2}} \sim \bar{U}\gamma^\mu\gamma^5 D \\
v^{4\mu} &\equiv A^{4\mu} \sim \bar{U}\gamma^\mu U + \bar{D}\gamma^\mu D,
\end{aligned}$$

for the vector mesons, and

$$
\begin{aligned}
x_{UU}^{\mu} &\equiv \frac{A^{10\mu} + iA^{11\mu} + A^{12\mu} + iA^{13\mu}}{2} \sim U^T C \gamma^{\mu} \gamma^5 U, \\
x_{DD}^{\mu} &\equiv \frac{A^{10\mu} + iA^{11\mu} - A^{12\mu} - iA^{13\mu}}{2} \sim D^T C \gamma^{\mu} \gamma^5 D, \quad (3.29) \\
x_{UD}^{\mu} &\equiv \frac{A^{14\mu} + iA^{15\mu}}{\sqrt{2}} \sim D^T C \gamma^{\mu} \gamma^5 U, \\
s_{UD}^{\mu} &\equiv \frac{A^{6\mu} - iA^{5\mu}}{\sqrt{2}} \sim U^T C \gamma^{\mu} D,
\end{aligned}
$$

for the vector baryons.

There are different approaches on how to introduce vector mesons at the effective Lagrangian level. At the tree level they are all equivalent.

Based on this premise, the minimal kinetic Lagrangian is:

$$
\mathscr{L}_{\text{kinetic}} = -\frac{1}{2} \text{Tr}\left[\widetilde{W}_{\mu\nu} \widetilde{W}^{\mu\nu} \right] - \frac{1}{4} B_{\mu\nu} B^{\mu\nu} - \frac{1}{2} \text{Tr}\left[F_{\mu\nu} F^{\mu\nu} \right] + m^2 \text{Tr}\left[C_{\mu} C^{\mu} \right],
$$
$$(3.30)$$

where $\widetilde{W}_{\mu\nu}$ and $B_{\mu\nu}$ are the ordinary field strength tensors for the electroweak gauge fields. Strictly speaking the terms above are not only kinetic ones since the Lagrangian contains a mass term as well as self interactions. The tilde on W^a indicates that the associated states are not yet the SM weak triplets: in fact these states mix with the composite vectors to form mass eigenstates corresponding to the ordinary W and Z bosons. $F_{\mu\nu}$ is the field strength tensor for the new $SU(4)$ vector bosons,

$$
F_{\mu\nu} = \partial_{\mu} A_{\nu} - \partial_{\nu} A_{\mu} - i\tilde{g} \left[A_{\mu}, A_{\nu} \right], \quad (3.31)
$$

and the vector field C_{μ} is defined by

$$
C_{\mu} \equiv A_{\mu} - \frac{g}{\tilde{g}} G_{\mu}. \quad (3.32)
$$

and G_{μ} is given by

$$
G_{\mu} = g W_{\mu}^a L^a + g' B_{\mu} Y, \quad (3.33)
$$

where L^a and Y are the generators of the left-handed and hypercharge transformations, as defined in Appendix A.3, with Y. The parameter $\tilde{g}$ represents the coupling among the vectors and the ratio $\frac{g}{\tilde{g}}$ is phenomenologically very important because it sets the mixing among gauge eigenstates and composite vectors eigenstates. The mass term in Eq. (3.30) is gauge invariant, and gives a degenerate mass to all composite vector bosons, while leaving the actual gauge bosons massless. (The latter acquires mass as usual from the covariant derivative term of the scalar matrix M, after spontaneous symmetry breaking.)

The C_μ fields couple with M via gauge invariant operators. Up to dimension four operators the Lagrangian is

$$
\mathcal{L}_{M-C} = \tilde{g}^2\, r_1 \text{Tr}\left[C_\mu C^\mu M M^\dagger \right] + \tilde{g}^2\, r_2\, \text{Tr}\left[C_\mu M C^{\mu T} M^\dagger \right]
$$
$$
+ i\, \tilde{g}\frac{r_3}{2}\, \text{Tr}\left[C_\mu \left(M (D^\mu M)^\dagger - (D^\mu M) M^\dagger \right) \right]
$$
$$
+ \tilde{g}^2\, s\, \text{Tr}\left[C_\mu C^\mu \right] \text{Tr}\left[M M^\dagger \right].
\tag{3.34}
$$

The dimensionless parameters r_1, r_2, r_3, s parameterize the strength of the interactions between the composite scalars and vectors in units of $\tilde{g}$, and are therefore naturally expected to be of order one. However, notice that for $r_1 = r_2 = r_3 = 0$ the overall Lagrangian possesses two independent $SU(2)_L \times U(1)_R \times U(1)_V$ global symmetries. One for the terms involving M and one for the terms involving C_μ.[2] The Higgs potential only breaks the symmetry associated with M, while leaving the symmetry in the vector sector unbroken. This *enhanced symmetry* guarantees that all r-terms are still zero after loop corrections. Moreover if one chooses r_1, r_2, r_3 to be small the near enhanced symmetry will protect these values against large corrections [14, 15].

3.3.3 Fermions and Yukawa Interactions

The fermionic content of the effective theory consists of the SM quarks and leptons, the new lepton doublet $L = (N, E)$ introduced to cure the Witten anomaly, and a composite techniquark-technigluon doublet.

We now consider the limit according to which the $SU(4)$ symmetry is, at first, extended to ordinary quarks and leptons. Of course, we will need to break this symmetry to accommodate the SM phenomenology. We start by arranging the $SU(2)$ doublets in $SU(4)$ multiplets as we did for the techniquarks in Eq. (3.5). We therefore introduce the four component vectors q^i and l^i,

$$
q^i = \begin{pmatrix} u_L^i \\ d_L^i \\ -i\sigma^2 u_R^{i\,*} \\ -i\sigma^2 d_R^{i\,*} \end{pmatrix}, \quad
l^i = \begin{pmatrix} v_L^i \\ e_L^i \\ -i\sigma^2 v_R^{i\,*} \\ -i\sigma^2 e_R^{i\,*} \end{pmatrix},
\tag{3.35}
$$

where i is the generation index. Note that such an extended $SU(4)$ symmetry automatically predicts the presence of a right handed neutrino for each generation. In addition to the SM fields there is an $SU(4)$ multiplet for the new leptons,

---

[2] The gauge fields explicitly break the original $SU(4)$ global symmetry to $SU(2)_L \times U(1)_R \times U(1)_V$, where $U(1)_R$ is the T^3 part of $SU(2)_R$, in the $SU(2)_L \times SU(2)_R \times U(1)_V$ subgroup of $SU(4)$.

$$L = \begin{pmatrix} N_L \\ E_L \\ -i\sigma^2 N_R^* \\ -i\sigma^2 E_R^* \end{pmatrix}, \tag{3.36}$$

and a multiplet for the techniquark-technigluon bound state,

$$\tilde{Q} = \begin{pmatrix} \tilde{U}_L \\ \tilde{D}_L \\ -i\sigma^2 \tilde{U}_R^* \\ -i\sigma^2 \tilde{D}_R^* \end{pmatrix}. \tag{3.37}$$

The techniquark-technigluon states, $\tilde{Q}$, being bound states of the underlying MWT model, have a dynamical mass.

With this arrangement, the electroweak covariant derivative for the fermion fields can be written

$$D_\mu = \partial_\mu - i\,g\,G_\mu(Y_V)\,, \tag{3.38}$$

where $Y_V = 1/3$ for the quarks, $Y_V = -1$ for the leptons, $Y_V = -3y$ for the new lepton doublet, and $Y_V = y$ for the techniquark-technigluon bound state. Based on this matter content, we write the following gauge part of the fermion Lagrangian:

$$\begin{aligned} \mathscr{L}_{\text{fermion}} = {}& i\,\bar{q}_{\dot\alpha}^i \bar{\sigma}^{\mu,\dot\alpha\beta} D_\mu q_\beta^i + i\,\bar{l}_{\dot\alpha}^i \bar{\sigma}^{\mu,\dot\alpha\beta} D_\mu l_\beta^i \\ &+ i\,\overline{L}_{\dot\alpha} \bar{\sigma}^{\mu,\dot\alpha\beta} D_\mu L_\beta + i\,\overline{\tilde{Q}}_{\dot\alpha} \bar{\sigma}^{\mu,\dot\alpha\beta} D_\mu \tilde{Q}_\beta \\ &+ x\,\overline{\tilde{Q}}_{\dot\alpha} \bar{\sigma}^{\mu,\dot\alpha\beta} C_\mu \tilde{Q}_\beta. \end{aligned} \tag{3.39}$$

We now turn to the issue of providing masses to the SM fermions. In the first chapter the simplest ETC model has been briefly reviewed. Many extensions of TC have been suggested in the literature to address this problem. Some of the extensions use another strongly coupled gauge dynamics, others introduce fundamental scalars. Many variants of the schemes presented above exist and a review of the major models is the one by Hill and Simmons [16]. At the moment there is not yet a consensus on which is the correct ETC. In our phenomenological approach will we parameterize our ignorance about a complete ETC theory by simply coupling the fermions to our low energy effective Higgs throughout the ordinary effective SM Yukawa interactions and we assume that any dangerous FCNC operator is strongly suppressed and therefore negligible.

3.3.4 Phenomenological Use of the Modified Weinberg Sum Rules

In order to make contact with the underlying gauge theory, and discriminate between different classes of models, we make use of the modified Weinberg sum rules discussed already in the first chapter. In [4] it was argued that the zeroth Weinberg sum rule—which is nothing but the definition of the S parameter

$$S = 4\pi \left[\frac{F_V^2}{M_V^2} - \frac{F_A^2}{M_A^2} \right], \tag{3.40}$$

and the first sum rule,

$$F_V^2 - F_A^2 = F_\pi^2, \tag{3.41}$$

do not receive significant contributions from the near conformal region, and are therefore unaffected. In these equations M_V (M_A) and F_V (F_A) are mass and decay constant of the vector-vector (axial-vector) meson, respectively, in the limit of zero electroweak gauge couplings. F_π is the decay constant of the pions: since this is a model of dynamical electroweak symmetry breaking, $F_\pi = 246\,\text{GeV}$. The heavy vector boson masses are:

$$M_V^2 = m^2 + \frac{\tilde{g}^2 \, (s - r_2) \, v^2}{4},$$
$$M_A^2 = m^2 + \frac{\tilde{g}^2 \, (s + r_2) \, v^2}{4}, \tag{3.42}$$

and

$$F_V = \frac{\sqrt{2} M_V}{\tilde{g}},$$
$$F_A = \frac{\sqrt{2} M_A}{\tilde{g}} \chi,$$
$$F_\pi^2 = (1 + 2\omega) F_V^2 - F_A^2, \tag{3.43}$$

where

$$\omega \equiv \frac{v^2 \tilde{g}^2}{4M_V^2}(1 + r_2 - r_3), \quad \chi \equiv 1 - \frac{v^2 \, \tilde{g}^2 \, r_3}{4M_A^2}. \tag{3.44}$$

Then Eqs. (3.40) and (3.41) give

$$S = \frac{8\pi}{\tilde{g}^2} \left(1 - \chi^2\right), \tag{3.45}$$
$$r_2 = r_3 - 1. \tag{3.46}$$

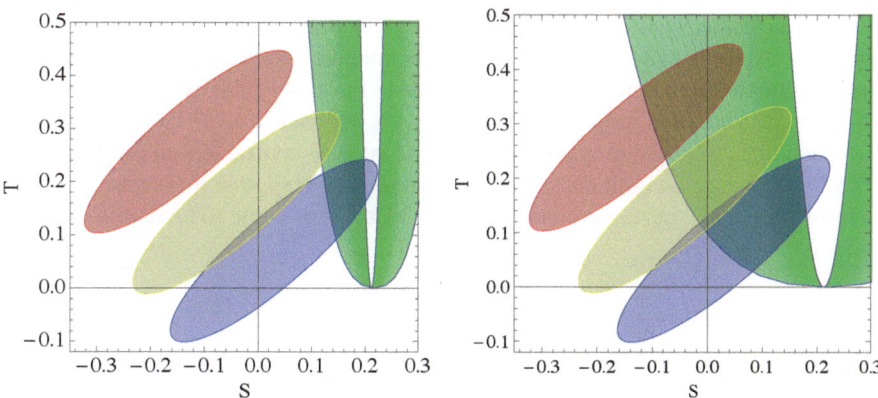

Fig. 3.2 The ellipses represent the 90 % confidence region for the S and T parameters. The ellipses, from lower to higher, are obtained for a reference Higgs mass of 117, 300 GeV, and 1 TeV, respectively. The contribution from the TC sector of the MWT theory per se and from the new leptons is expressed by the green region. The *left* panel has been obtained using a SM type hypercharge assignment while the *right* one is for $y = 1$

The second sum rule, corresponding to a zero on the right hand side of the following equation, does receive important contributions from the near conformal region, and is modified to

$$F_V^2 M_V^2 - F_A^2 M_A^2 = a \frac{8\pi^2}{d(R)} F_\pi^4, \tag{3.47}$$

where a is expected to be positive and $\mathcal{O}(1)$, and $d(R)$ is the dimension of the representation of the underlying fermions [4]. For each of these sum rules a more general spectrum would involve a sum over all the vector and axial states.

In the effective Lagrangian we codify the walking behavior in a being positive and $\mathcal{O}(1)$, and the minimality of the theory in S being small. A small S is both due to the small number of flavors in the underlying theory and to the near conformal dynamics, which reduces the contribution to S relative to a running theory [4].

3.3.5 Passing the Electroweak Precision Tests

We have studied the effects of the lepton family on the electroweak parameters in [13], we summarize here the main results in Fig. 3.2. The ellipses represent the 90 % confidence region for the S and T parameters. The ellipses, from lower to higher, are obtained for a reference Higgs mass of 117 300 GeV, and 1 TeV, respectively. The contribution from the MWT theory per se and of the new leptons [17] is expressed by the green region. The left panel has been obtained using a SM type hypercharge assignment while the right one is for $y = 1$. In both pictures the regions of overlap between the theory and the precision contours are achieved when the upper component of the weak isospin doublet is lighter than the lower component. The opposite case leads to a total S which is larger than the one predicted within the new strongly coupled dynamics per se. This is due to the sign of the hypercharge for the new leptons. The mass range used in the plots is $M_Z \leqslant m_{E,N} \leqslant 10\, M_Z$. The plots have been obtained assuming a Dirac mass for the new neutral lepton (in the case of a SM hypercharge assignment). The analysis for the Majorana mass case has been performed in [18] where one can again show that it is possible to be within the 90 % contours.

3.3.6 The Next to Minimal Walking Technicolor Theory (NMWT)

The theory with three technicolors contains an even number of electroweak doublets, and hence it is not subject to a Witten anomaly. The doublet of technifermions, is then represented again as:

$$Q_L^{\{C_1,C_2\}} = \begin{pmatrix} U^{\{C_1,C_2\}} \\ D^{\{C_1,C_2\}} \end{pmatrix}_L, \quad Q_R^{\{C_1,C_2\}} = \left(U_R^{\{C_1,C_2\}},\ D_R^{\{C_1,C_2\}} \right). \tag{3.48}$$

Here $C_i = 1, 2, 3$ is the technicolor index and $Q_{L(R)}$ is a doublet (singlet) with respect to the weak interactions. Since the two-index symmetric representation of $SU(3)$ is complex the flavor symmetry is $SU(2)_L \times SU(2)_R \times U(1)$. Only three Goldstones emerge and are absorbed in the longitudinal components of the weak vector bosons.

Gauge anomalies are absent with the choice $Y = 0$ for the hypercharge of the left-handed technifermions:

$$Q_L^{(Q)} = \begin{pmatrix} U^{(+1/2)} \\ D^{(-1/2)} \end{pmatrix}_L. \tag{3.49}$$

Consistency requires for the right-handed technifermions (isospin singlets):

$$Q_R^{(Q)} = \left(U_R^{(+1/2)},\ D_R^{-1/2} \right),$$
$$Y = +1/2, -1/2. \tag{3.50}$$

All of these states will be bound into hadrons. There is no need for an associated fourth family of leptons, and hence it is not expected to be observed in the experiments.

Here the low-lying technibaryons are fermions constructed with three techniquarks in the following way:

$$B_{f_1,f_2,f_3;\alpha} = Q_{L;\alpha,f_1}^{\{C_1,C_2\}} Q_{L;\beta,f_2}^{\{C_3,C_4\}} Q_{L;\gamma,f_3}^{\{C_5,C_6\}} \varepsilon^{\beta\gamma} \varepsilon_{C_1 C_3 C_5} \varepsilon_{C_2 C_4 C_6}. \tag{3.51}$$

where $f_i = 1, 2$ corresponds to U and D flavors, and we are not specifying the flavor symmetrization which in any event will have to be such that the full technibaryon wave function is fully antisymmetrized in technicolor, flavor and spin. α, β, and γ assume the values of one or two and represent the ordinary spin. Similarly we can construct different technibaryons using only right-handed fields or a mixture of left- and right-handed ones.

This model has also been recently investigated on the lattice and found to break chiral symmetry [19].

3.4 Beyond Minimal Technicolor

When going beyond MWT one finds new and interesting theories able to break the electroweak symmetry while featuring a walking dynamics and yet not at odds with precision measurements, at least when comparing with the naive S parameter. A compendium of these theories can be found in [20]. Here we will review only the principal type of models one can construct.

3.4.1 Partially Gauged Technicolor

A small modification of the traditional TC approach, which neither involves additional particle species nor more complicated gauge groups, allows constructing several other viable candidates. It consists in letting only one doublet of techniquarks transform non-trivially under the electroweak symmetries with the rest being electroweak singlets, as first suggested in [13] and later also used in [21]. Still, all techniquarks transform under the TC gauge group. Thereby only one techniquark doublet contributes directly[3] to the oblique parameter which is thus kept to a minimum for theories which need more than one family of techniquarks to be quasi-conformal. It is the condensation of that first electroweakly charged family that breaks the electroweak symmetry. The techniquarks which are uncharged under the electroweak gauge group are natural building blocks for components of DM.

3.4.2 Split Technicolor

We summarize here also another possibility [13] according to which we keep the technifermions gauged under the electroweak theory in the fundamental representation of the $SU(N)$ TC group while still reducing the number of techniflavors needed

---

[3] Via TC interactions all of the matter content of the theory will affect physical observables associated to the sector coupled to the electroweak symmetry.

to be near the conformal window. Like for the partially gauged case described above this can be achieved by adding matter uncharged under the weak interactions. The difference to Sect. 3.4.1 is that this part of matter transforms under a different representation of the TC gauge group than the part coupled directly to the electroweak sector. For example, for definiteness let's choose it to be a massless Weyl fermion in the adjoint representation of the TC gauge group. The resulting theory has the same matter content as N_f-flavor super QCD but without the scalars; hence the name *Split Technicolor*. The matter content of *Split Technicolor* lies between that of super QCD and QCD-like theories with matter in the fundamental representation. We note that a split TC-like theory has been used in [22], to investigate the strong CP problem.

In [23] one can find an explicit example of (near) conformal TC with two types of technifermions, i.e. transforming according to two different representations of the underlying TC gauge group [20, 24]. The model possesses a number of interesting properties to recommend it over the earlier models of dynamical electroweak symmetry breaking:

- Features the lowest possible value of the naive S parameter [25, 26] while possessing a dynamics which, if not jumping, is walking.
- Contains, overall, the lowest possible number of fermions.
- Yields natural DM candidates.

Due to the above properties we term this model *Ultra Minimal near conformal Technicolor* (UMT). It is constituted by an $SU(2)$ technicolor gauge group with two Dirac flavors in the fundamental representation also carrying electroweak charges, as well as, two additional Weyl fermions in the adjoint representation but singlets under the SM gauge groups. Recently the $SU(2)$ dynamics with two Dirac flavors in the fundamental representation was investigated on the lattice in [27]. Here it was shown that the pattern of chiral symmetry breaking is indeed $SU(4)$ breaking to $Sp(4)$.

By arranging the additional fermions in higher dimensional representations, it is possible to construct models which have a particle content smaller than the one of partially gauged technicolor theories. In fact instead of considering additional fundamental flavors we shall consider adjoint flavors. Note that for two colors there exists only one distinct two-indexed representation.

3.5 Vanilla Technicolor

Despite the different envisioned underlying gauge dynamics it is a fact that the SM structure alone requires the extensions to contain, at least, the following chiral symmetry breaking pattern (insisting on keeping the custodial symmetry of the SM):

$$SU(2)_\text{L} \times SU(2)_\text{R} \rightarrow SU(2)_\text{V} . \tag{3.52}$$

We will call this common sector of any technicolor extension of the SM, the *vanilla* sector. The reason for such a name is that the vanilla sector is common to old models of technicolor featuring running and walking dynamics. It is worth mentioning that the *vanilla* sector is common not only to technicolor extensions but to several extensions, even of extra-dimensional type, in which the Higgs sector can be viewed as composite. In fact, the effective Lagrangian we are about to introduce can be used for modeling several extensions with a common vanilla sector respecting the same constraints spelled out in [28]. The natural candidate for a walking technicolor model featuring exactly this global symmetry is NMWT [6].

Based on the *vanilla symmetry* breaking pattern we describe the low energy spectrum in terms of the lightest spin one vector and axial-vector iso-triplets $V^{\pm,0}$, $A^{\pm,0}$ as well as the lightest iso-singlet scalar resonance H. In QCD the equivalent states are the $\rho^{\pm,0}$, $a_1^{\pm,0}$ and the $f_0(600)$ [29]. It has been argued in [3, 30], using Large N arguments, and in [13, 20], using the saturation of the trace of the energy momentum tensor, that models of dynamical electroweak symmetry breaking featuring (near) conformal dynamics contain a composite Higgs state which is light with respect to the new strongly coupled scale ($4\pi v$ with $v \simeq 246\,\text{GeV}$). These indications have led to the construction of models of technicolor with a naturally light composite Higgs. Recent investigations using Schwinger-Dyson [31] and gauge-gravity dualities [32] also arrived to the conclusion that the composite Higgs can be light.[4] The three technipions $\Pi^{\pm,0}$ produced in the symmetry breaking become the longitudinal components of the W and Z bosons.

The composite spin one and spin zero states and their interaction with the SM fields are described via the following effective Lagrangian in which we developed, first for minimal models of walking technicolor [15, 28]:

$$
\begin{aligned}
\mathscr{L}_{\text{boson}} = {} & -\frac{1}{2}\text{Tr}\left[\widetilde{W}_{\mu\nu}\widetilde{W}^{\mu\nu}\right] - \frac{1}{4}\widetilde{B}_{\mu\nu}\widetilde{B}^{\mu\nu} - \frac{1}{2}\text{Tr}\left[F_{L\mu\nu}F_L^{\mu\nu} + F_{R\mu\nu}F_R^{\mu\nu}\right] \\
& + m^2\,\text{Tr}\left[C_{L\mu}^2 + C_{R\mu}^2\right] + \frac{1}{2}\text{Tr}\left[D_\mu M D^\mu M^\dagger\right] - \tilde{g}^2\,r_2\,\text{Tr}\left[C_{L\mu}M C_R^\mu M^\dagger\right] \\
& - \frac{i\,\tilde{g}\,r_3}{4}\text{Tr}\left[C_{L\mu}\left(M D^\mu M^\dagger - D^\mu M M^\dagger\right) + C_{R\mu}\left(M^\dagger D^\mu M - D^\mu M^\dagger M\right)\right] \\
& + \frac{\tilde{g}^2 s}{4}\text{Tr}\left[C_{L\mu}^2 + C_{R\mu}^2\right]\text{Tr}\left[M M^\dagger\right] + \frac{\mu^2}{2}\text{Tr}\left[M M^\dagger\right] - \frac{\lambda}{4}\text{Tr}\left[M M^\dagger\right]^2,
\end{aligned}
\tag{3.53}
$$

where $\widetilde{W}_{\mu\nu}$ and $\widetilde{B}_{\mu\nu}$ are the ordinary electroweak field strength tensors, $F_{L/R\mu\nu}$ are the field strength tensors associated to the vector meson fields $A_{L/R\mu}$,[5] and the $C_{L\mu}$ and $C_{R\mu}$ fields are

$$
C_{L\mu} \equiv A_{L\mu} - \frac{g}{\tilde{g}}\widetilde{W}_\mu, \qquad C_{R\mu} \equiv A_{R\mu} - \frac{g'}{\tilde{g}}\widetilde{B}_\mu.
\tag{3.54}
$$

---

[4] The Higgs boson here is identified with the lightest 0^{++} state of the theory saturating the trace of the energy momentum tensor of the theory.

[5] In [28], where the chiral symmetry is $SU(4)$, there is an additional term whose coefficient is labeled r_1. With an $SU(N) \times SU(N)$ chiral symmetry this term is just identical to the s term.

The 2×2 matrix M is

$$M = \frac{1}{\sqrt{2}} \left[v + H + 2 i \, \pi^a \, T^a \right], \quad a = 1, 2, 3 \tag{3.55}$$

where π^a are the Goldstone bosons produced in the chiral symmetry breaking, $v = \mu/\sqrt{\lambda}$ is the corresponding VEV, H is the composite Higgs, and $T^a = \sigma^a/2$, where σ^a are the Pauli matrices. The covariant derivative is

$$D_\mu M = \partial_\mu M - i g \tilde{W}_\mu^a T^a M + i g' M \, \tilde{B}_\mu T^3. \tag{3.56}$$

When M acquires a VEV, the Lagrangian of Eq. (3.53) contains mixing matrices for the spin one fields. The mass eigenstates are the ordinary SM bosons, and two triplets of heavy mesons, of which the lighter (heavier) ones are denoted by $R_1^\pm$ ($R_2^\pm$) and R_1^0 (R_2^0). These heavy mesons are the only new particles, at low energy, relative to the SM.

Now we must couple the SM fermions. The interactions with the Higgs and the spin one mesons are mediated by an unknown ETC sector, and can be parametrized at low energy by Yukawa terms, and mixing terms with the C_L and C_R fields. Assuming that the ETC interactions preserve parity and do not generate extra flavor violation beyond the SM like Yukawa terms, the most general form for the quark Lagrangian is[6]

$$\mathscr{L}_{\text{quark}} = \bar{q}_L^i \, i \slashed{D} q_{iL} + \bar{q}_R^i \, i \slashed{D} q_{iR}$$
$$- \left[\bar{q}_L^i \, (Y_u)_i^j \, M \, \frac{1 + \tau^3}{2} \, q_{jR} + \bar{q}_L^i \, (Y_d)_i^j \, M \, \frac{1 - \tau^3}{2} \, q_{jR} + \text{h.c.} \right], \tag{3.57}$$

where i and j are generation indices, $i = 1, 2, 3$, $q_{iL/R}$ are electroweak doublets, Y_u and Y_d are 3×3 complex matrices. The covariant derivatives are the ordinary electroweak ones,

$$\slashed{D} q_{iL} = \left(\slashed{\partial} - i g \, \tilde{W}^a \, T^a - i g' \tilde{B} Y_L \right) q_{iL},$$
$$\slashed{D} q_{iR} = \left(\slashed{\partial} - i g' \tilde{B} Y_R \right) q_{iR}, \tag{3.58}$$

where $Y_L = 1/6$ and $Y_R = \text{diag}(2/3, -1/3)$. One can exploit the global symmetries of the kinetic terms to reduce the number of physical parameters in the Yukawa matrices. Thus we can take

$$Y_u = \text{diag}(y_u, y_c, y_t), \quad Y_d = V \, \text{diag}(y_d, y_s, y_b), \tag{3.59}$$

and

---

[6] The lepton sector works out in a similar way, the only difference being the possible presence of Majorana neutrinos.

$$q_L^i = \begin{pmatrix} u_{iL} \\ V_i^j d_{jL} \end{pmatrix}, \quad q_R^i = \begin{pmatrix} u_{iR} \\ d_{iR} \end{pmatrix}, \tag{3.60}$$

where V is the CKM matrix.

It is possible to further reduce the number of independent couplings using the Weinberg sum rules discussed above. For example in NMWT, featuring technifermions with three technicolors transforming according to the two-index symmetric representation of the technicolor gauge group, the naive one-loop S parameter is $S = 1/\pi \simeq 0.3$: this is a reasonable input for S in Eq. (3.40).

With $S = 0.3$ the remaining parameters are $M_A, \tilde{g}, s$ and M_H, with s and M_H having a sizable effect in processes involving the composite Higgs.[7]

3.6 WW: Scattering in Technicolor and Unitarity

The simplest argument often used to predict the existence of yet undiscovered particles at the TeV scale comes from unitarity of longitudinal gauge boson scattering amplitudes. If the electroweak symmetry breaking sector (EWSB) is weakly interacting, unitarity implies that new particle states must show up below one TeV, being these spin zero isosinglets (the Higgs boson) or spin one isotriplets (e.g. Kaluza-Klein modes). A strongly interacting EWSB sector can however change this picture, because of the strong coupling between the pions (eaten by the longitudinal components of the SM gauge bosons) and the other bound states of the strongly interacting sector. An illuminating example comes from QCD. In [33] it was shown that for six colors or more, the 770 GeV ρ meson is enough to delay the onset of unitarity violation of the pion-pion scattering amplitude up to well beyond 1 GeV. Here the 't Hooft large N limit was used, however an even lower number of colors is needed to reach a similar delay of unitarity violation when an alternative large N limit is used [34]. Scaling up to the electroweak scale, this translates in a 1.5 TeV technivector being able to delay unitarity violation of longitudinal gauge boson scattering amplitudes up to 4 TeV or more. As we discussed in the previous sections such a model, however, would not be realistic for other reasons: a large contribution to the S parameter [25], and large FCNC if the ordinary fermions acquire mass via an old fashioned ETC, to mention the most relevant ones. It is therefore interesting to analyze the pion-pion scattering in generic models of walking technicolor. We follow the analyses performed in [35, 36].

In the effective theory for technicolor the scattering amplitudes for the longitudinal SM gauge bosons approach at large energies the scattering amplitudes for the corresponding eaten pions. We mainly analyze the contribution to the $\pi\pi$ scattering amplitude from a spin zero isosinglet and a spin one isotriplet, and consider the case in which a spin two isosinglet contributes as well.

---

[7] The information on the spectrum alone is not sufficient to constrain s, but it can be measured studying other physical processes.

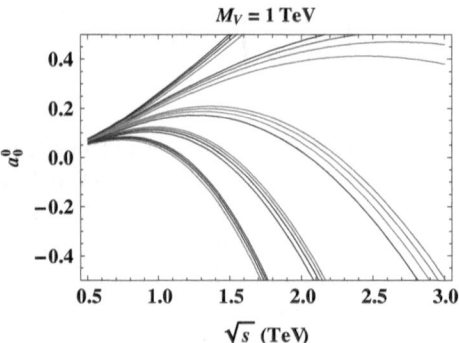

Fig. 3.3 $I = 0 \, J = 0$ partial wave amplitude for the $\pi\pi$ scattering. Here a Higgs with mass $M_H = 200 \, \text{GeV}$, and a spin-one vector meson with mass $M_V = 1 \, \text{TeV}$ contribute to the full amplitude. The different groups of curves correspond, from top to bottom, to $g_{V\pi\pi} = 2, 2.5, 3, 3.5, 4$. The different *curves* within each group correspond, from *top* to *bottom*, to $h = 0, 0.1, 0.15, 0.2$. Nonzero values of $g_{V\pi\pi}$ and h give negative contributions to the linear term in s in the amplitude, and may lead to a delay of unitarity violation

3.6.1 Spin Zero + Spin One

The isospin invariant amplitude for the pion-pion elastic scattering is [37]:

$$A(s, t, u) = \left(\frac{1}{F_\pi^2} - \frac{3 g_{V\pi\pi}^2}{M_V^2} \right) s - \frac{h^2}{M_H^2} \frac{s^2}{s - M_H^2} - g_{V\pi\pi}^2 \left[\frac{s - u}{t - M_V^2} + \frac{s - t}{u - M_V^2} \right]. \tag{3.61}$$

Note that our normalization for $g_{V\pi\pi}$, which is the heavy vector to two-pions effective coupling, differs by a factor of $\sqrt{2}$ from that of Ref. [37]. The scalar H contribution is proportional to the coupling h. These couplings are simply related to the ones of the Vanilla technicolor Lagrangian, but the specific relation is not relevant here.

The amplitude of Eq. (3.61) has an s-channel pole in the Higgs exchange. In the vicinity of this pole the propagator should be modified to include the Higgs width. In order to catch the essential features of the unitarization process we will take the Higgs to be a relatively narrow state, and consider values of $\sqrt{s}$ far away from M_H, where the finite width effects can be neglected. If the Higgs or any other state is not sufficiently narrow to be treated at the tree level, it would be relevant to investigate the effects due to unitarity corrections using specific unitarization schemes as done for example in [38]. In order to study unitarity of the $\pi\pi$ scattering the most general amplitude should be expanded in its isospin I and spin J components, a_J^I. However the $I = 0 \, J = 0$ component,

$$a_0^0(s) = \frac{1}{64\pi} \int_{-1}^{1} d\cos\theta \, [3A(s, t, u) + A(t, s, u) + A(u, t, s)], \tag{3.62}$$

has the worst high energy behavior, and is therefore sufficient for our analysis. Since we are interested in testing unitarity at few TeVs in presence of a light Higgs, we set $M_H = 200 \, \text{GeV}$ as a reference value, and study the regions in the $(M_V, g_{V\pi\pi})$ plane in which a_0^0 is unitary up to 3 TeV, for different values of h. If the Higgs mass is larger than 200 GeV but still smaller than or of the same size of M_V, we expect our results to be qualitatively similar, even though finite width effects might be important due to the pole in the s-channel. If the Higgs mass is much larger than M_V the theory is Higgsless at low energies. This case was studied in Ref. [35], and applies also to the light Higgs scenario if H is decoupled from the pions, i.e. $h = 0$.

In order to study the effect of the Higgs exchange on the scattering amplitude, consider the high energy behavior of $A(s, t, u)$,

$$A(s, t, u) \sim \left(\frac{1}{F_\pi^2} - \frac{3g_{V\pi\pi}^2}{M_V^2} - \frac{h^2}{M_H^2} \right) s \,. \tag{3.63}$$

This shows that the Higgs exchange provides an additional negative contribution at large energies, which, together with the vector meson, contributes to delay unitarity violation to higher energies. In Fig. 3.3 a_0^0 is plotted as a function of $\sqrt{s}$ for $M_V = 1 \, \text{TeV}$, $M_H = 200 \, \text{GeV}$, and different values of $g_{V\pi\pi}$ and h. The different groups of curves from top to bottom correspond to $g_{V\pi\pi} = 2, 2.5, 3, 3.5$, and 4. For comparison, the QCD value that follows from $\Gamma(\rho \to \pi\pi) \simeq 150 \, \text{MeV}$ would be $g_{V\pi\pi} \simeq 5.6$.[8] Within each group, the top curve corresponds to the Higgsless case, $h = 0$, while the remaining ones correspond, from top to bottom, to $h = 0.1, 0.15$, and 0.2. For small values of $g_{V\pi\pi}$ the presence of a light Higgs delays unitarity violation to higher energies: if the partial wave amplitude has a maximum near 0.5 the delay is dramatic.

For a given value of M_V, the presence of a light Higgs enlarges the interval of values of $g_{V\pi\pi}$ for which the theory is unitary, provided that $|h|$ is not too large.

3.6.2 Spin Zero + Spin One + Spin Two

In addition to spin-zero and spin-one mesons, the low energy spectrum can contain spin two mesons as well [37]. The contribution of a spin-two meson F_2 to the invariant amplitude is

$$A_2(s, t, u) = \frac{g_2^2}{2(M_{F_2}^2 - s)} \left[-\frac{s^2}{3} + \frac{t^2 + u^2}{2} \right] - \frac{g_2^2 s^3}{12 M_{F_2}^4} \,, \tag{3.64}$$

where M_{F_2} and g_2 are mass and coupling with the pions, respectively. A reference value for g_2 can be obtained from QCD: $m_{f_2} \simeq 1275 \, \text{MeV}$ and $\Gamma(f_2 \to$

---

[8] Figure 3.3 does not reproduce a scaled up version of QCD $\pi\pi$ scattering. For the latter to occur, the vector resonance should be as large as $(246 \, \text{GeV}/93 \, \text{MeV}) \times 770 \, \text{MeV} \simeq 2 \, \text{TeV}$. However in a theory with walking dynamics the resonances are expected to be lighter than in a running setup.

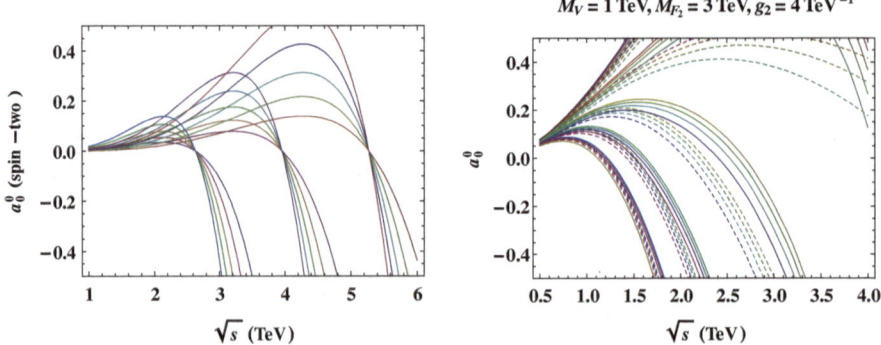

Fig. 3.4 *Left* contribution from the spin-two exchanges to the $I = 0\, J = 0$ partial wave amplitude of the $\pi\pi$ scattering. The different groups of *curves* correspond, from *left* to *right*, to $M_{F_2} = 2, 3, 4\,\text{TeV}$. Within each group, the different *curves* correspond, from smaller to wider, to $g_2 = 2, 2.5, 3, 3.5, 4\,\text{TeV}^{-1}$. *Right* $I = 0\, J = 0$ partial wave amplitude with all channels included (spin-zero, -one, and -two). The *dashed curves* reproduce Fig. 3.3, with just the spin-zero and the spin-one channels included. The *solid curves* contain also the spin-two exchanges, for $M_{F_2} = 3\,\text{TeV}$, and $g_2 = 4\,\text{TeV}^{-1}$. If unitarity is violated at negative values of a_0^0, the spin-two exchanges may lead to a delay of unitarity violation

$\pi\pi) \simeq 160\,\text{MeV}$ give $|g_2| \simeq 13\,\text{GeV}^{-1}$ so that $|g_2|F_\pi \simeq 1.2$. Scaling up to the eletroweak scale results in $|g_2| \simeq 4\,\text{TeV}^{-1}$. The contribution of F_2 to the $I = 0$ $J = 0$ partial wave amplitude is given in Fig. 3.4 (left) for different values of M_{F_2} and g_2. Notice that the amplitude is initially positive, and then becomes negative at large values of $\sqrt{s}$. If M_{F_2} is large enough, the positive contribution can balance the negative contribution from the spin-zero and spin-one channels, shown in Fig. 3.3. This can lead to a further delay of unitarity violation, as shown in Fig. 3.4 (right). Here the curves of Fig. 3.3 are redrawn dashed, while the full contribution from spin-zero, spin-one, and spin-two is shown by the solid lines, for $M_{F_2} = 3\,\text{TeV}$ and $g_2 = 4\,\text{TeV}^{-1}$. If unitarity is violated at negative values of a_0^0, then the spin-two contribution delays the violation to higher energies.

The unitarity analysis presented here is for generic Vanilla technicolor theories, or any other model, featuring spin zero, one, and two resonances. The specialization to running and walking technicolor is described in detail in [35, 36]. The bottom line is that it is possible to delay the onset of unitarity violation, at the effective Lagrangian level, for phenomenologically viable values of the couplings and masses of the composite spectrum.

3.6.3 Introducing and Constraining Custodial Technicolor

We now constrain also models proposed in [15, 39] which, at the effective Lagrangian level, possess an explicit *custodial* symmetry for the S parameter. We will refer to this class of models as custodial technicolor (CT) [40] . The new custodial symmetry is present in the BESS models [14, 41, 42] which will therefore be constrained as well. In this case we expect our constraints to be similar to the ones also discussed in [43].

Custodial technicolor corresponds to the case for which $M_A = M_V = M$ and $\chi = 0$. The effective Lagrangian acquires a new symmetry, relating a vector and an axial field, which can be interpreted as a custodial symmetry for the S parameter [15, 39]. The only non-zero parameters are now:

$$W = \frac{g^2}{\tilde{g}^2} \frac{M_W^2}{M^2} , \tag{3.65}$$

$$Y = \frac{g'^2}{2\tilde{g}^2} \frac{M_W^2}{M^2} (2 + 4y^2) . \tag{3.66}$$

A CT model cannot be achieved in walking dynamics and must be interpreted as a new framework. In other words CT does not respect the Weinberg sum rules and hence it can only be considered as a phenomenological type model in search of a fundamental strongly coupled theory. To make our point clearer note that a degenerate spectrum of light spin-one resonances (i.e. $M < 4\pi F_\pi$) leads to a very large $\hat{S} = g^2 F_\pi^2 / 4M^2$. We needed only the first sum rule together with the statement of degeneracy of the spectrum to derive this $\hat{S}$ parameter. This statement is universal and it is true for WT and ordinary technicolor. The Eichten and Lane [44] scenario of almost degenerate and very light spin-one states can only be achieved within a near CT models. A very light vector meson with a small number of techniflavors fully gauged under the electroweak can be accommodated in CT. This scenario was considered in [45, 46] and our constraints apply here.

We find that in CT it is possible to have a very light and degenerate spin-one spectrum if $\tilde{g}$ is sufficiently large, of the order say of 8 or larger as in the WT case.

We constrained the electroweak parameters intrinsic to WT or CT, however, in general other sectors may contribute to the electroweak observables, an explicit example is the new heavy lepton family introduced above [13].

To summarize we have suggested in [40] a way to constrain WT theories with any given S parameter. We have further constrained relevant models featuring a custodial symmetry protecting the S parameter. When increasing the value of the S parameter while reducing the amount of walking we recover the technicolor constraints [25]. We found bounds on the lightest spectrum of WT and CT theories with an intrinsically small S parameter. Our results are applicable to *any* dynamical model of electroweak symmetry breaking featuring near conformal dynamics á la walking technicolor.

3.6.4 An ETC Example for MWT Versus Top Mass

It is instructive to present a simple model [47] which shows how one can embed MWT in an extended technicolor model capable of generating the top quark mass.

When the techni-quarks are not in the fundamental representation of the technicolor group it can be hard to feed down the electroweak symmetry breaking condensate to generate the SM fermion masses [21, 24]. Here, following [47], we wish to highlight that the minimal model can be recast as an SO(3) theory with fundamental representation techni-quarks. The model can therefore rather easily be enlarged to an extended technicolor theory [24] in the spirit of many examples in the literature. We will concentrate on the top quark sector—the ETC gauge bosons in this sector violate weak isospin and one must be careful to compute their contribution to the T parameter [48].

We start by recognizing that adjoint multiplets of SU(2) can be written as fundamental representations of SO(3). This trick will now allow us to enact a standard ETC pattern from the literature—it is particularly interesting that for this model of the higher dimensional representation techniquarks there is a simple ETC model. We will follow the path proposed in [49] where we gauge the full flavor symmetry of the fermions.

If we were simply interested in the fourth family then the enlarged ETC symmetry is a Pati-Salam type unification. We stack the doublets

$$\left[\begin{pmatrix} U^a \\ D^a \end{pmatrix}_L, \ \begin{pmatrix} N \\ E \end{pmatrix}_L\right], \quad [U_R^a, \ N_R], \quad [D_R^a, \ E_R] \tag{3.67}$$

into 4 dimensional multiplets of SU(4). One then invokes some symmetry breaking mechanism at an ETC scale (we will not speculate on the mechanism here though see Fig. 3.1)

$$SU(4)_{ETC} \to SO(3)_{TC} \times U(1)_Y \tag{3.68}$$

The technicolor dynamics then proceeds to generate a techniquark condensate $\langle \bar{U}U \rangle = \langle \bar{D}D \rangle \neq 0$. The massive gauge bosons associated with the broken ETC generators can then feed the symmetry breaking condensate down to generate fourth family lepton masses

$$m_N = m_E \simeq \frac{\langle \bar{U}U \rangle}{\Lambda_{ETC}^2} \tag{3.69}$$

One could now naturally proceed to include the third (second, first) family by raising the ETC symmetry group to SU(8) (SU(12), SU(16)) and a series of appropriate symmetry breakings. This would generate masses for all the SM fermions but no isospin breaking mass contributions within fermion doublets. The simplest route to generate such splitting is to make the ETC group chiral so that different ETC couplings determine the isospin $+1/2$ and $-1/2$ masses. Let us only enforce such a pattern for the top quark and fourth family here since the higher ETC scales are far beyond experimental probing.

We can, for example, have the SU(7) multiplets

$$\left[\begin{pmatrix} U^a \\ D^a \end{pmatrix}_L , \ \begin{pmatrix} N \\ E \end{pmatrix}_L , \ \begin{pmatrix} t^c \\ b^c \end{pmatrix}_L \right], \qquad \left[U^a_R , \ N_R , \ t^c_R \right] \qquad (3.70)$$

here a will become the technicolor index and c the QCD index. We also have a right handed SU(4) ETC group that only acts on

$$\left[D^a_R, E_R \right] \qquad (3.71)$$

The right handed bottom quark is left out of the ETC dynamics and only has proto-QCD SU(3) dynamics. The bottom quark will thus be left massless. The symmetry breaking scheme at, for example, a single ETC scale would then be

$$SU(7) \times SU(4) \times SU(3) \rightarrow SO(3)_{TC} \times SU(3)_{QCD} \qquad (3.72)$$

The top quark now also acquires a mass from the broken gauge generators naively equal to the fourth family lepton multiplet. Walking dynamics has many features though that one would expect to overcome the traditional small size of the top mass in ETC models. Firstly the enhancement of the techniquark self energy at high momentum enhances the ETC generated masses by a factor potentially as large as $\Lambda_{ETC}/\Sigma(0)$.

The technicolor coupling is near conformal and strong so the ETC dynamics will itself be quite strong at its breaking scale which will tend to enhance light fermion masses [50]. In this ETC model the top quark will also feel the effects of the extra massive octet of axial gluon-like gauge fields that may induce a degree of top condensation a là top color models [51]. We conclude that a 4–8 TeV ETC scale for generating the top mass is possible. In this model the fourth family lepton would then have a mass of the same order and well in excess of the current search limit $M_Z/2$.

References

1. J.R. Andersen, O. Antipin, G. Azuelos, L. Del Debbio, E. Del Nobile, S. Di Chiara, T. Hapola M. Jarvinen et al. Eur. Phys. J. Plus **126**, 81 (2011). [arXiv:1104.1255 [hep-ph]]
2. K. Yamawaki, M. Bando K. Matumoto, Phys. Rev. Lett. **56**, 1335 (1986)
3. D.K. Hong, S.D.H. Hsu, F. Sannino, Phys. Lett. B **597**, 89 (2004). [arXiv:hep-ph/0406200]
4. T. Appelquist, F. Sannino, Phys. Rev. D **59**, 067702 (1999). [arXiv:hep-ph/9806409]
5. R. Sundrum, S.D.H. Hsu, Nucl. Phys. B **391**, 127 (1993). [arXiv:hep-ph/9206225]
6. F. Sannino, K. Tuominen, Phys. Rev. D **71**, 051901 (2005). [arXiv:hep-ph/0405209]
7. T.A. Ryttov, F. Sannino, Phys. Rev. D **78**, 065001 (2008). [arXiv:0711.3745 [hep-th]]
8. S. Catterall, F. Sannino, Phys. Rev. D **76**, 034504 (2007). [arXiv:0705.1664 [hep-lat]]

9. A.J. Hietanen, K. Rummukainen, K. Tuominen, Phys. Rev. D **80**, 094504 (2009). [arXiv:0904.0864 [hep-lat]]
10. L. Del Debbio, A. Patella, C. Pica, Phys. Rev. D **81**, 094503 (2010). [arXiv:0805.2058 [hep-lat]]
11. S. Catterall, J. Giedt, F. Sannino, J. Schneible, JHEP **0811**, 009 (2008). [arXiv:0807.0792 [hep-lat]]
12. E. Witten, Phys. Lett. B **117**, 324 (1982)
13. D.D. Dietrich, F. Sannino, K. Tuominen, Phys. Rev. D **72**, 055001 (2005). [arXiv:hep-ph/0505059]
14. R. Casalbuoni, A. Deandrea, S. De Curtis, D. Dominici, R. Gatto, M. Grazzini, Phys. Rev. D **53**, 5201 (1996). [arXiv:hep-ph/9510431]
15. T. Appelquist, P.S. Rodrigues da Silva, F. Sannino, Phys. Rev. D **60**, 116007 (1999). [arXiv:hep-ph/9906555]
16. C.T. Hill, E.H. Simmons, Phys. Rep. **381**, 235 (2003). [Erratum-ibid. 390, 553 (2004)] [arXiv:hep-ph/0203079]
17. H.-J. He, N. Polonsky, S.-f. Su, Phys. Rev. D **64**, 053004 (2001). [hep-ph/0102144]
18. K. Kainulainen, K. Tuominen, J. Virkajarvi, Phys. Rev. D **75**, 085003 (2007). [arXiv:hep-ph/0612247]
19. Z. Fodor, K. Holland, J. Kuti, D. Nogradi, C. Schroeder, C.H. Wong, arXiv:1205.1878 [hep-lat]
20. D.D. Dietrich, F. Sannino, Phys. Rev. D **75**, 085018 (2007). [arXiv:hep-ph/0611341]
21. N.D. Christensen, R. Shrock, Phys. Lett. B **632**, 92 (2006). [arXiv:hep-ph/0509109]
22. S.D.H. Hsu, F. Sannino, Phys. Lett. B **605**, 369 (2005). [arXiv:hep-ph/0408319]
23. T.A. Ryttov, F. Sannino, Phys. Rev. D **78**, 115010 (2008). [arXiv:0809.0713 [hep-ph]]
24. K.D. Lane, E. Eichten, Phys. Lett. B **222**, 274 (1989)
25. M.E. Peskin, T. Takeuchi, Phys. Rev. Lett. **65**, 964 (1990)
26. M.E. Peskin, T. Takeuchi, Phys. Rev. D **46**, 381 (1992)
27. R. Lewis, C. Pica, F. Sannino, Phys. Rev. D **85**, 014504 (2012). [arXiv:1109.3513 [hep-ph]]
28. R. Foadi, M.T. Frandsen, T.A. Ryttov, F. Sannino, Phys. Rev. D **76**, 055005 (2007). [arXiv:0706.1696 [hep-ph]]
29. A. Belyaev, R. Foadi, M.T. Frandsen, M. Jarvinen, F. Sannino, A. Pukhov, Phys. Rev. D **79**, 035006 (2009). [arXiv:0809.0793 [hep-ph]]
30. F. Sannino, arXiv:0804.0182 [hep-ph]
31. A. Doff, A.A. Natale, P.S. Rodrigues da Silva, Phys. Rev. D **77**, 075012 (2008). [arXiv:0802.1898 [hep-ph]]
32. M. Fabbrichesi, M. Piai, L. Vecchi, Dynamical electro-weak symmetry breaking from deformed AdS: vector mesons. Phys. Rev. D **78**, 045009 (2008). [arXiv:0804.0124 [hep-ph]]
33. M. Harada, F. Sannino, J. Schechter, Phys. Rev. D **69**, 034005 (2004). [arXiv:hep-ph/0309206]
34. F. Sannino, J. Schechter, Phys. Rev. D **76**, 014014 (2007). [arXiv:0704.0602 [hep-ph]]
35. R. Foadi, F. Sannino, Phys. Rev. D **78**, 037701 (2008). [arXiv:0801.0663 [hep-ph]]
36. R. Foadi, M. Jarvinen, F. Sannino, Phys. Rev. D **79**, 035010 (2009). [arXiv:0811.3719 [hep-ph]]
37. M. Harada, F. Sannino, J. Schechter, Phys. Rev. D **54**, 1991 (1996). [arXiv:hep-ph/9511335]
38. D. Black, A.H. Fariborz, S. Moussa, S. Nasri, J. Schechter, Phys. Rev. D **64**, 014031 (2001). [hep-ph/0012278]
39. Z.y. Duan, P.S. Rodrigues da Silva, F. Sannino, Nucl. Phys. B **592**, 371 (2001) [arXiv:hep-ph/0001303]
40. R. Foadi, M.T. Frandsen, F. Sannino, Phys. Rev. D **77**, 097702 (2008). [hep-ph]]
41. R. Casalbuoni, S. De Curtis, D. Dominici, F. Feruglio, R. Gatto, Int. J. Mod. Phys. A **4**, 1065 (1989)
42. R. Casalbuoni, A. Deandrea, S. De Curtis, D. Dominici, F. Feruglio, R. Gatto, M. Grazzini, Phys. Lett. B **349**, 533 (1995). [arXiv:hep-ph/9502247]

43. R. Casalbuoni, F. Coradeschi, S. De Curtis, D. Dominici, Phys. Rev. D **77**, 095005 (2008). [arXiv:0710.3057 [hep-ph]]
44. E. Eichten, K. Lane, Phys. Lett. B **669**, 235 (2008). [arXiv:0706.2339 [hep-ph]]
45. A.R. Zerwekh, Eur. Phys. J. C **46**, 791 (2006). [arXiv:hep-ph/0512261]
46. A.R. Zerwekh, C.O. Dib, R. Rosenfeld, Phys. Rev. D **75**, 097702 (2007). [arXiv:hep-ph/0702167]
47. N. Evans, F. Sannino, arXiv:hep-ph/0512080
48. R.S. Chivukula, B.A. Dobrescu, J. Terning, Phys. Lett. B **353**, 289 (1995). [arXiv:hep-ph/9503203]
49. L. Randall, Nucl. Phys. B **403**, 122 (1993). [arXiv:hep-ph/9210231]
50. N.J. Evans, Phys. Lett. B **331**, 378 (1994). [arXiv:hep-ph/9403318]
51. V.A. Miransky, T. Nonoyama, K. Yamawaki, Mod. Phys. Lett. A **4**, 1409 (1989)

Chapter 4
Composite Dark Matters: Coda

Abstract The dark side holds the 96 % share of the Universe. Given that to describe the remaining 4 % we need at least three forces it is very likely that the dark side of the Universe hides many new forces. We will suggest few possibilities yielding candidates for composite dark matter and inflation making use of new strong dynamics.

4.1 Composite Dark Matter

Experimental observations strongly indicate that the universe is flat and predominantly made of unknown forms of matter. Defining with Ω the ratio of the density to the critical density, observations indicate that the fraction of matter amounts to $\Omega_{\text{matter}} \sim 0.3$ of which the normal baryonic one is only $\Omega_{\text{baryonic}} \sim 0.044$. The amount of non-baryonic matter is termed dark matter. The total Ω in the universe is dominated by dark matter and pure energy (dark energy) with the latter giving a contribution $\Omega_\Lambda \sim 0.7$ (see for example [1, 2]). Most of the dark matter is "cold" (i.e. non-relativistic at freeze-out) and significant fractions of hot dark matter are hardly compatible with data. What constitutes dark matter is a question relevant for particle physics and cosmology. A WIMP (Weakly Interacting Massive Particles) can be the dominant part of the non-baryonic dark matter contribution to the total Ω. Axions can also be dark matter candidates but only if their mass and couplings are tuned for this purpose. It would be theoretically very pleasing to be able to relate the dark matter and the baryon energy densities in order to explain the ratio $\Omega_{\text{DM}}/\Omega_B \sim 5$ [3]. We know that the amount of baryons in the universe $\Omega_B \sim 0.04$ is determined solely by the cosmic baryon asymmetry $n_B/n_\gamma \sim 6 \times 10^{-10}$. This is so since the baryon–antibaryon annihilation cross section is so large, that virtually all antibaryons annihilate away, and only the contribution proportional to the asymmetry remains. This asymmetry can be dynamically generated after inflation. We do not know, however, if the dark matter density is determined by thermal freeze-out, by an asymmetry, or by something else. Thermal freeze-out needs a $\sigma v \approx 3 \; 10^{-26} \, \text{cm}^3/\text{s}$

F. Sannino, *Dynamical Stabilization of the Fermi Scale*, SpringerBriefs in Physics,   89
DOI: 10.1007/978-3-642-33341-5_4, © The Author(s) 2013

which is of the electroweak scale, suggesting a dark matter mass in the TeV range. If Ω_{DM} is determined by thermal freeze-out, its proximity to Ω_B is just a fortuitous coincidence and is left unexplained.

If instead $\Omega_{DM} \sim \Omega_B$ is not accidental, then the theoretical challenge is to define a consistent scenario in which the two energy densities are related. Since Ω_B is a result of an asymmetry, then relating the amount of dark matter to the amount of baryon matter can very well imply that Ω_{DM} is related to the same asymmetry that determines Ω_B. Such a condition is straightforwardly realized if the asymmetry for the dark matter particles is fed in by the non-perturbative electroweak sphaleron transitions, that at temperatures much larger than the temperature T_* of the electroweak phase transition equilibrates the baryon, lepton and dark matter asymmetries. Implementing this condition implies the following requirements:

1. Dark matter must be (or must be a composite state of) a fermion, chiral (and thereby non-singlet) under the weak $SU(2)_L$, and carrying an anomalous (quasi)-conserved quantum number B'.
2. Dark matter (or its constituents) must have an annihilation cross section much larger than electroweak $\sigma_{ann} \gg 3\,10^{-26} cm^3/s$, to ensure that Ω_{DM} is determined dominantly by the B' asymmetry.

The first condition ensures that a global quantum number corresponding to a linear combination of B, L and B' has a weak anomaly, and thus dark matter carrying B' charge is produced in anomalous processes together with left-handed quarks and leptons [4, 5]. At temperatures $T \gg T_*$ electroweak anomalous processes are in thermal equilibrium, and equilibrate the various asymmetries $Y_{\Delta B} = c_L Y_{\Delta L} = c_{B'} Y_{\Delta B'} \sim \mathcal{O}(10^{-10})$. Here the Y_Δ's represent the difference in particle number densities $n - \bar{n}$ normalized to the entropy density s, e.g. $Y_{\Delta B} = (n_B - \bar{n}_B)/s$. These are convenient quantities since they are conserved during the Universe thermal evolution.

At $T \gg M_{DM}$ all particle masses can be neglected, and c_L and $c_{B'}$ are order one coefficients, determined via chemical equilibrium conditions enforced by elementary reactions faster than the Universe expansion rate [6]. These coefficients can be computed in terms of the particle content, finding e.g. $c_L = -28/51$ in the standard model and $c_L = -8/15$ in the Minimal Supersymmetric Standard Model.

At $T \ll M_{DM}$, the B' asymmetry gets suppressed by a Boltzmann exponential factor $e^{-M_{DM}/T}$. A key feature of sphaleron transitions is that their rate gets suddenly suppressed at some temperature T_* slightly below the critical temperature at which $SU(2)_L$ starts to be spontaneously broken. Thereby, if $M_{DM} < T_*$ the B' asymmetry gets frozen at a value of $\mathcal{O}(Y_{\Delta B})$, while if instead $M_{DM} > T_*$ it gets exponentially suppressed as $Y_{\Delta B'}/Y_{\Delta B} \sim e^{-M_{DM}/T}$.

More in detail, the sphaleron processes relate the asymmetries of the various fermionic species with chiral electroweak interactions as follows. If B', B and L are the only quantum numbers involved then the relation is:

$$\frac{Y_{\Delta B'}}{Y_{\Delta B}} = c \cdot \mathscr{S}\left(\frac{M_{DM}}{T_*}\right), \qquad c = \bar{c}_{B'} + \bar{c}_L \frac{Y_{\Delta L}}{Y_{\Delta B}}, \qquad (4.1)$$

where the order-one $\bar{c}_{L,B'}$ coefficients are related to the $c_{L,B'}$ above in a simple way. The explicit numerical values of these coefficients depend also on the order of the finite temperature electroweak phase transition via the imposition or not of the weak isospin charge neutrality. In [7, 8] the dependence on the order of the electroweak phase transition was studied in two explicit models, and it was found that in all cases the coefficients remain of order one. The statistical function $\mathscr{S}$ is:

$$\mathscr{S}(z) = \begin{cases} \frac{6}{4\pi^2} \int_0^\infty dx \, x^2 \cosh^{-2}\left(\frac{1}{2}\sqrt{x^2+z^2}\right) & \text{for fermions}, \\ \frac{6}{4\pi^2} \int_0^\infty dx \, x^2 \sinh^{-2}\left(\frac{1}{2}\sqrt{x^2+z^2}\right) & \text{for bosons}. \end{cases} \qquad (4.2)$$

with $S(0) = 1(2)$ for bosons (fermions) and $S(z) \simeq 12 \, (z/2\pi)^{3/2} e^{-z}$ at $z \gg 1$. We assumed the standard model fields to be relativistic and checked that this is a good approximation even for the top quark [7, 8]. The statistic function leads to the two limiting results:

$$\frac{Y_{\Delta B'}}{Y_{\Delta B}} = c \times \begin{cases} \mathscr{S}(0) & \text{for } M_{\text{DM}} \ll T_* \\ 12 \, (M_{\text{DM}}/2\pi T_*)^{3/2} \, e^{-M_{\text{DM}}/T_*} & \text{for } M_{\text{DM}} \gg T_* \end{cases}. \qquad (4.3)$$

Under the assumption that all antiparticles carrying B and B' charges are annihilated away we have $Y_{\Delta B'}/Y_{\Delta B} = n_{B'}/n_B$. The observed dark matter density

$$\frac{\Omega_{\text{DM}}}{\Omega_B} = \frac{M_{\text{DM}} \, n_{B'}}{m_p \, n_B} \approx 5 \qquad (4.4)$$

(where $m_p \approx 1\,\text{GeV}$) can be reproduced for two possible values of the dark matter mass:

i) $M_{\text{DM}} \sim 5\,\text{GeV}$ if $M_{\text{DM}} \ll T_*$, times model dependent order one coefficients.
ii) $M_{\text{DM}} \approx 8\,T_* \approx 2\,\text{TeV}$ if $M_{\text{DM}} \gg T_*$, with a mild dependence on the model-dependent order unity coefficients.

The first solution is well known [4] and corresponds to a light dark matter candidate. While the second condition would lead to a dark matter candidate with a mass of the order of the electroweak scale. This is the asymmetric dark matter paradigm. Many related properties (valid also for symmetric type scenarios) are not yet constrained by our current knowledge of dark matter, for example the specific dark matter candidate may or may not be a stable particle [9] and it may or may not be identified with its antiparticle [10]. However, very recently it was shown that there is a neat way to *discover* the existence asymmetric dark matter by studying the cosmic sum rules introduced in [11, 12].

Asymmetric dark matter candidates were put forward in [10] as technibaryons, in [13] as Goldstone bosons, and subsequently in many diverse forms [7, 14–19]. There is also the possibility of mixed dark matter [20], i.e. having both a thermally-produced symmetric component and an asymmetric one.

4.2 Dark Matter in Technicolor

If dark matter is an elementary particle, the asymmetric scenario needs dark matter to be a chiral fermion with $SU(2)_L$ interactions, which is very problematic. Bounds from direct detection are violated. Furthermore, a Yukawa coupling λ of dark matter to the Higgs gives the desired dark matter mass $M_{DM} \sim \lambda v \sim 2\,\mathrm{TeV}$ if $\lambda \sim 4\pi$ is non-perturbative, hinting to a dynamically generated mass associated to some new strongly interacting dynamics [5, 7, 8, 10]. This assumption also solves the problem with direct detection bounds, which are satisfied if dark matter is a composite $SU(2)_L$-singlet state, made of elementary fermions charged under $SU(2)_L$.

This can be realized by introducing a strongly-interacting ad-hoc 'hidden' gauge group. A more interesting identification comes from Technicolor. In such a scenario, dark matter would be the lightest (quasi)-stable composite state carrying a B' charge of a theory of dynamical electroweak breaking featuring a spectrum of technibaryons (B') and technipions (Π). The TIMP (Technicolor Interacting Massive Particle) can have a number of phenomenologically interesting properties.

i) A traditional TIMP mass can be approximated by $m_{B'} = M_{DM} \approx n_Q \Lambda_{TC}$ where n_Q is the number of techniquarks Q bounded into B' and Λ_{TC} is the constituent mass, so that $M_{DM}/m_p \approx n_Q \Lambda_{TC}/3\Lambda_{QCD}$. Denoting by f_π (F_Π) the (techni)pion decay constant, we have $F_\Pi/f_\pi = \sqrt{D/3}\,\Lambda_{TC}/\Lambda_{QCD}$ where D_Q is the dimension of the constituent fermions representation ($D = 3$ in QCD).[1] Finally, the electroweak breaking order parameter is obtained as $v^2 = N_D F_\Pi^2$, from the sum of the contribution of the N_D electroweak techni-doublets. Putting all together yields the estimate:

$$M_{DM} \approx n \frac{n_Q}{\sqrt{3 D_Q N_D}} \frac{v}{f_\pi} m_p = 2.2\,\mathrm{TeV} \qquad (4.5)$$

where the numerical value corresponds to the smallest number of constituents and of techniquarks $n_Q = D_Q = 2$ and $N_D = 1$.

ii) A generic dynamical origin of the breaking of the electroweak symmetry can lead to several natural interesting dark matter candidates (see [21] for a list of relevant references). A very interesting case is the one in which the TIMP is a pseudo-Goldstone boson [7, 8]. In this case one can observe these states also at colliders experiments [14].

According to [22] the sphaleron contribution to the technibaryon decay rate is negligible because exponentially suppressed, unless the technibaryon is heavier than several TeV.

---

[1] The large-N counting relevant for a generic extension of TC type can be found in Appendix F of [21].

Grand unified theories (GUTs) suggest that the baryon number B is violated by dimension-6 operators suppressed by the GUT scale $M_{\text{GUT}} \sim 2 \cdot 10^{16}$ GeV, yielding a proton life-time [23]

$$\tau(p \to \pi^0 e^+) \sim \frac{M_{\text{GUT}}^4}{m_p^5} \sim 10^{41} \text{ s.} \qquad (4.6)$$

If B' is similarly violated at the same high scale M_{GUT}, our TIMP would decay with life-time

$$\tau \sim \frac{M_{\text{GUT}}^4}{M_{\text{DM}}^5} \sim 10^{26} \text{ s,} \qquad (4.7)$$

which falls in the ball-park required by the phenomenological analysis to explain the PAMELA anomaly [9]. Models of unification of the standard model couplings in the presence of a dynamical electroweak symmetry breaking mechanism have been recently explored [24, 25]. Interestingly, the scale of unification suggested by the phenomenological analysis emerges quite naturally [25].

Low energy TIMP and nucleon (quasi)-stability imply that, in the primeval Universe, at temperatures $T \lesssim M_{\text{GUT}}$ perturbative violation of the B' and B global charges is strongly suppressed. Since this temperature is presumably larger than the reheating temperature, it is unlikely that Ω_B and Ω_{DM} result directly from an asymmetry generated in B' or B. More likely, the initial seed yielding Ω_{DM} and Ω_B could be an initial asymmetry in lepton number L that, much along the lines of well studied leptogenesis scenarios [26], feeds the B and B' asymmetries through the sphaleron effects.Indeed, it has been shown that it is possible to embed seesaw-types of scenarios in theories of dynamical symmetry breaking,while keeping the scale of the L-violating Majorana masses as low as $\sim 10^3$ TeV [27]. In Minimal Walking Technicolor [28, 29], one additional (technisinglet) SU(2)-doublet must be introduced to cancel the odd-number-of-doublets anomaly [30]. An asymmetry in the L' global charge associated with these new states can also serve as a seed for the B and B' asymmetries. In [31] it has been shown that is possible to embed a low energy see-saw mechanism for the fourth family Leptons in the Minimal Walking Technicolor extension of the standard model.

Assuming that TIMP decays is dominantly due to effective four-fermion operators, its decay modes significantly depend on the technicolor gauge group. In the following L generically denotes any standard model fermion, quark or lepton, possibly allowed by the Lorentz and gauge symmetries of the theory.

- If the technicolor group is SU(3), the situation is analogous to ordinary QCD: the TIMP is a fermionic QQQ state, and effective $QQQL$ operators gives $TIMP \to \Pi^- \ell^+$ decays. This leads to hard leptons, but together with an excess of $\bar{p}$, from the $\Pi^- \to \bar{c}$ decay (in view of $\Pi^- \simeq W_L^-$).
- If the technicolor group is SU(4) the situation is that the TIMP is a bosonic $QQQQ$ state, and effective $QQQQ$ operators lead to its decay into techni-pions.

• Finally, if the technicolor gauge group is SU(2) the TIMP is a bosonic QQ state, (as put forward in [7]), and effective $QQLL$ operators lead to TIMP decays into two L. Since the fundamental representation of SU(2) is pseudoreal, one actually gets an interesting dynamics analyzed in detail in [7]. Here the TIMP is a pseudo-Goldstone boson of the underlying gauge theory.

An SU(2) technicolor model compatible with the desired features is obtained assuming that the left component of the Dirac field Q has zero hypercharge and is a doublet under SU(2)$_L$, so that the TIMP is a scalar QQ with no overall weak interactions, and the four-fermion operator $(QQ)\partial_\mu(\bar{L}\gamma_\mu L)$ allows it to decay. Such operator is possible for both standard model leptons and quarks, so that the TIMP branching ratios into $\ell^+\ell^-$ and $q\bar{q}$ is a free parameter.

4.2.1 Current Experimental Status

From the experimental point of view null results from several experiments, such as CDMS [32] and Xenon10/100 [33, 34], have placed stringent constraints on WIMP-nucleon cross sections. Interestingly DAMA [35] and CoGeNT [36] have both produced evidence for an annual modulation signature for dark matter, as expected due to the relative motion of the Earth with respect to the dark matter halo. These results support a light WIMP with mass of order a few GeV, which offers the attractive possibility of a common mechanism for baryogenesis and dark matter production. At first glance it seems that the WIMP-nucleon cross sections required by DAMA and CoGeNT have been excluded by CDMS and Xenon upon assuming spin-independent interactions between WIMPs and nuclei (with protons and neutrons coupling similarly to WIMPs), however a number of resolutions for this puzzle have been proposed in the literature [37–43]. Interestingly, also recent results from the CRESST-II experiment report signals of light dark matter [44].

A composite origin of dark matter, along the lines detailed above, is therefore quite an intriguing possibility given that the bright side of the universe, constituted mostly by nucleons, is also composite. Thus a new strongly-coupled theory could be at the heart of dark matter. Furthermore for the first time on the lattice, a technicolor-type extension of the standard model, expected to naturally yield a light dark matter candidate, as introduced in [7] and used in [42] to reconcile the experimental observations, has been investigated [45]. Here it was shown that strongly interacting theories can, indeed, support electroweak symmetry breaking while yielding natural light dark matter candidates.

Models of dynamical breaking of the electroweak symmetry do support the possibility of generating the experimentally observed baryon (and possibly also the technibaryon/dark matter) asymmetry of the universe directly at the electroweak phase transition [46–48]. Electroweak baryogenesis [49] is, however, impossible in the standard model [50].

4.3 Composite Inflation

Another prominent physics problem is inflation [51–56], the mechanism responsible for an early rapid expansion of our universe. Inflation, similar to the standard model Higgs mechanism, is also modeled traditionally via the introduction of new scalar fields. However, field theories featuring fundamental scalars are unnatural. The reason being that typically these theories lead to the introduction of symmetry-unprotected super-renormalizable operators, such as the scalar quadratic mass operator. Quantum corrections, therefore, introduce untamed divergencies which have to be fine-tuned away. Following the composite Higgs section [17, 57] one can imagine a new natural strong dynamics underlying the inflationary mechanism [58]. In [59] we spelled out the setup for generic models of composite inflation. Another logical possibility is that theories with scalars are gauge-dual to theories featuring only fermionic degrees of freedom [60–63].

We briefly review here the general setup for strongly coupled inflation [58, 59]. We start by identifying the inflaton with one of the lightest composite states of a generic strongly coupled theory and denote it with Φ. This state has mass dimension d. This is the physical dimension coming from the sum of the engineering dimensions of the elementary fields constituting the inflaton augmented by the anomalous dimensions due to quantum corrections in the underlying gauge theory. We concentrate [58] on the non-Goldstone sector of the theory.[2]

We then consider the following coupling to gravity in the Jordan frame:

$$\mathscr{S}_{CI,J} = \int d^4x \sqrt{-g} \left[-\frac{\mathscr{M}^2 + \xi \Phi^{\frac{2}{d}}}{2} g^{\mu\nu} R_{\mu\nu} + \mathscr{L}_{\Phi} \right],$$

$$\mathscr{L}_{\Phi} = g^{\mu\nu} \Phi^{\frac{2-2d}{d}} \partial_\mu \Phi \partial_\nu \Phi - V(\Phi), \tag{4.8}$$

with $\mathscr{L}_{\Phi}$ the low energy effective Lagrangian for the field Φ constrained by the symmetries of the underlying strongly coupled theory. In this framework $\mathscr{M}$ is not automatically the Planck constant M_{Pl}. The non-minimal coupling to gravity is controlled by the dimensionless coupling ξ. The non-analytic power of Φ emerges because we are requiring a dimensionless coupling with the Ricci scalar. Abandoning the conformality requirement allows for operators with integer powers of Φ when coupling to the Ricci scalar. However a new energy scale must be introduced to match the mass dimensions. We diagonalize the gravity-composite dynamics model via the conformal transformation:

$$g_{\mu\nu} \to \tilde{g}_{\mu\nu} = \Omega(\Phi)^2 g_{\mu\nu}, \quad \Omega(\Phi)^2 = \frac{\mathscr{M}^2 + \xi \Phi^{\frac{2}{d}}}{M_p^2}, \tag{4.9}$$

---

[2] The Goldstone sector, if any, associated to the potential dynamical spontaneous breaking of some global symmetries of the underlying gauge theory will be investigated elsewhere.

such that

$$\tilde{g}^{\mu\nu} = \Omega^{-2} g^{\mu\nu}, \quad \sqrt{-\tilde{g}} = \Omega^4 \sqrt{-g}. \tag{4.10}$$

We use both the Palatini and the metric formulation. The difference between the two formulations resides in the fact that in the Palatini formulation the connection Γ is assumed not to be directly associated with the metric $g_{\mu\nu}$. Hence the Ricci tensor $R_{\mu\nu}$ does not transform under the conformal transformation. Applying the conformal transformation we arrive at the Einstein frame action:

$$\mathscr{S}_{CI,E} = \int d^4x \sqrt{-g} \left[-\frac{1}{2} M_p^2 \, g^{\mu\nu} R_{\mu\nu} + \Omega^{-2} \left(\Phi^{\frac{2-2d}{d}} + f \cdot 3M_p^2 \Omega'^2 \right) g^{\mu\nu} \partial_\mu \Phi \partial_\nu \Phi - \Omega^{-4} V(\Phi) \right]. \tag{4.11}$$

Primes denote derivatives with respect to Φ and tildes are dropped for convenience. $f = 1$ signifies the metric formulation [64–67] and $f = 0$ the Palatini one [68].

We introduce a canonically normalized field χ related to Φ via

$$\frac{1}{2} \tilde{g}^{\mu\nu} \partial_\mu \chi(\Phi) \partial_\nu \chi(\Phi) = \frac{1}{2} \left(\frac{d\chi}{d\Phi} \right)^2 \tilde{g}^{\mu\nu} \partial_\mu \Phi \partial_\nu \Phi, \tag{4.12}$$

with

$$\begin{aligned}
\frac{1}{2} \left(\frac{d\chi}{d\Phi} \right)^2 &= \Omega^{-2} \left(\Phi^{\frac{2-2d}{d}} + f \cdot 3M_p^2 \Omega'^2 \right) \\
&= \Omega^{-2} \left(1 + f \cdot \frac{3\xi^2}{d^2 M_p^2} \Omega^{-2} \Phi^{\frac{2}{d}} \right) \Phi^{\frac{2-2d}{d}}. \tag{4.13}
\end{aligned}$$

In terms of the canonically normalized field we have:

$$\mathscr{S}_{CI,E} = \int d^4x \sqrt{-g} \left[-\frac{1}{2} M_p^2 g^{\mu\nu} R_{\mu\nu} + \frac{1}{2} g^{\mu\nu} \partial_\mu \chi \partial_\nu \chi - U(\chi) \right]. \tag{4.14}$$

With

$$U(\chi) \equiv \Omega^{-4} V(\Phi). \tag{4.15}$$

Within this framework we determined in [59] useful expressions for the slow-roll parameters for composite inflation and provided the explicit example in which the inflaton emerges as the lightest glueball field associated to, in absence of gravity, a pure Yang-Mills theory. This theory constitutes the archetype of any composite model in flat space and consequently of models of composite inflation. We showed that it is possible to achieve successful glueball inflation. Furthermore the natural scale of compositeness associated to the underlying Yang-Mills gauge theory, for the consistency of the model, turns to be of the order of the grand unified scale. This result is in agreement with the scale of compositeness scale determined in [58] for a very different underlying model of composite inflation.

One can also show that, in the metric formulation, that unitarity-cutoff for inflaton–inflaton scattering is well above the energy scale relevant for composite inflation. It is now possible to envision a large number of new avenues to explore within this class of models.

4.4 Finale

Circa 96 % of the universe is made by unknown forms of matter and energy, while to describe the remaining 4 % one needs at least three fundamental forces, i.e. Quantum Electrodynamics (QED), Weak Interactions and QCD. Furthermore strong interactions are responsible for creating the bulk of the bright mass, i.e. the 4 %. It is therefore natural to expect that to correctly describe the rest of our universe while providing a sensible link to the visible component new forces will soon emerge. Here we have suggested three primary areas of research where new strong dynamics can occur. The first is the sector responsible for breaking spontaneously the electroweak symmetry. The standard model Higgs, or perhaps the entire standard model [61], could be replaced by new strongly interacting dynamics. We have also seen that another relevant physical application of new strong dynamics is in the construction of composite dark matter candidates with very interesting phenomenology. Last but not the least we have also envisioned the possibility that even the mechanism behind inflation can find its roots in new strong dynamics. It is therefore of the utmost importance to gain a deeper understanding of the nonperturbative regime of gauge theories of fundamental interactions. Quite excitingly we are only at the very beginning of the theoretical understanding and phenomenological impact and discovery of new strongly interacting theories.

References

1. J. Frieman, M. Turner, D. Huterer, Ann. Rev. Astron. Astrophys. **46**, 385 (2008), [arXiv:0803.0982 [astro-ph]]
2. J.A. Frieman, D. Huterer, E.V. Linder, M.S. Turner, Phys. Rev. **D67**, 083505 (2003), [arXiv:astro-ph/0208100]
3. E. Komatsu et al., [WMAP Collaboration], Astrophys. J. Suppl. **180**, 330 (2009), [arXiv:0803.0547 [astro-ph]]
4. D.B. Kaplan, Phys. Rev. Lett. **68**, 741 (1992)
5. S.M. Barr, R.S. Chivukula, E. Farhi, Phys. Lett. B **241**, 387 (1990)
6. J.A. Harvey, M.S. Turner, Phys. Rev. **D42**, 3344 (1990)
7. T.A. Ryttov, F. Sannino, Phys. Rev. **D78**, 115010 (2008), [arXiv:0809.0713 [hep-ph]]
8. S.B. Gudnason, C. Kouvaris, F. Sannino, Phys. Rev. **D74**, 095008 (2006), [arXiv:hep-ph/0608055]
9. E. Nardi, F. Sannino, A. Strumia, JCAP **0901**, 043 (2009), [arXiv:0811.4153 [hep-ph]]
10. S. Nussinov, Phys. Lett. B **165**, 55 (1985)
11. I. Masina, F. Sannino, JCAP **1109**, 021 (2011), [arXiv:1106.3353 [hep-ph]]

12. M.T. Frandsen, I. Masina, F. Sannino, Phys. Rev. **D83**, 127301 (2011), [arXiv:1011.0013 [hep-ph]]
13. S.B. Gudnason, C. Kouvaris, F. Sannino, Phys. Rev. **D73**, 115003 (2006), [arXiv:hep-ph/0603014]
14. R. Foadi, M.T. Frandsen, F. Sannino, Phys. Rev. **D80**, 037702 (2009), [arXiv:0812.3406 [hep-ph]]
15. M.Y. Khlopov, C. Kouvaris, Phys. Rev. **D78**, 065040 (2008), [arXiv:0806.1191 [astro-ph]]
16. D.D. Dietrich, F. Sannino, Phys. Rev. **D75**, 085018 (2007), [arXiv:hep-ph/0611341]
17. F. Sannino, Acta Phys. Polon. B **40**, 3533 (2009), [arXiv:0911.0931 [hep-ph]]
18. D.E. Kaplan, M.A. Luty, K.M. Zurek, Phys. Rev. **D79**, 115016 (2009), [arXiv:0901.4117 [hep-ph]]
19. M.T. Frandsen, F. Sannino, Phys. Rev. **D81**, 097704 (2010), [arXiv:0911.1570 [hep-ph]]
20. A. Belyaev, M.T. Frandsen, S. Sarkar, F. Sannino, Phys. Rev. **D83**, 015007 (2011), [arXiv:1007.4839 [hep-ph]]
21. F. Sannino, arXiv:0804.0182 (2008), [hep-ph]
22. V.A. Rubakov, Nucl. Phys. B **256**, 509 (1985)
23. J. Hisano, H. Murayama, T. Yanagida, Nucl. Phys. B **402**, 46 (1993), [arXiv:hep-ph/9207279]
24. N.D. Christensen, R. Shrock, Phys. Rev. **D72**, 035013 (2005), [arXiv:hep-ph/0506155]
25. S.B. Gudnason, T.A. Ryttov, F. Sannino, Phys. Rev. **D76**, 015005 (2007), [arXiv:hep-ph/0612230]
26. M. Fukugita, T. Yanagida, Phys. Lett. B **174**, 45 (1986)
27. T. Appelquist, R. Shrock, Phys. Lett. B **548**, 204 (2002), [arXiv:hep-ph/0204141]
28. D.D. Dietrich, F. Sannino, K. Tuominen, Phys. Rev. **D72**, 055001 (2005), [arXiv:hep-ph/0505059]
29. R. Foadi, M.T. Frandsen, T.A. Ryttov, F. Sannino, Phys. Rev. **D76**, 055005 (2007), [arXiv:0706.1696 [hep-ph]]
30. E. Witten, Phys. Lett. B **117**, 324 (1982)
31. M.T. Frandsen, I. Masina, F. Sannino, (2008), arXiv:0905.1331 [hep-ph]
32. Z. Ahmed et al., [CDMS-II Collaboration], Phys. Rev. Lett. **106**, 131302 (2011), [arXiv:1011.2482 [astro-ph.CO]]
33. J. Angle et al., [XENON10 Collaboration], Phys. Rev. Lett. **107**, 051301 (2011), [arXiv:1104.3088 [astro-ph.CO]]
34. E. Aprile et al., [XENON100 Collaboration], Phys. Rev. Lett. **107**, 131302 (2011), [arXiv:1104.2549 [astro-ph.CO]]
35. R. Bernabei et al., [DAMA Collaboration], Eur. Phys. J. C **56**, 333 (2008), [arXiv:0804.2741 [astro-ph]]
36. C.E. Aalseth et al., [CoGeNT Collaboration], Phys. Rev. Lett. **106**, 131301 (2011), [arXiv:1002.4703 [astro-ph.CO]]
37. M.Y. Khlopov, A.G. Mayorov, E.Y. Soldatov, Int. J. Mod. Phys. **D19**, 1385 (2010), [arXiv:1003.1144 [astro-ph.CO]]
38. D. Tucker-Smith, N. Weiner, Phys. Rev. **D64**, 043502 (2001), [arXiv:hep-ph/0101138]
39. S. Chang, J. Liu, A. Pierce, N. Weiner, I. Yavin, JCAP **1008**, 018 (2010), [arXiv:1004.0697 [hep-ph]]
40. J.L. Feng, J. Kumar, D. Marfatia, D. Sanford, Phys. Lett. B **703**, 124 (2011), [arXiv:1102.4331 [hep-ph]]
41. M.T. Frandsen, F. Kahlhoefer, J. March-Russell, C. McCabe, M. McCullough, K. Schmidt-Hoberg, Phys. Rev. **D84**, 041301 (2011), [arXiv:1105.3734 [hep-ph]]
42. E. Del Nobile, C. Kouvaris, F. Sannino, Phys. Rev. **D84**, 027301 (2011), [arXiv:1105.5431 [hep-ph]]
43. E. Del Nobile, C. Kouvaris, F. Sannino, J. Virkajarvi, (2012), arXiv:1111.1902 [hep-ph]
44. G. Angloher et al., (2012), arXiv:1109.0702 [astro-ph.CO]
45. R. Lewis, C. Pica, F. Sannino, Phys. Rev. **D85**, 014504 (2012), [arXiv:1109.3513 [hep-ph]]
46. J.M. Cline, M. Jarvinen, F. Sannino, Phys. Rev. **D78**, 075027 (2008), [arXiv:0808.1512 [hep-ph]]

47. M. Jarvinen, T.A. Ryttov, F. Sannino, Phys. Lett. B **680**, 251 (2009), [arXiv:0901.0496 [hep-ph]]
48. M. Jarvinen, T.A. Ryttov, F. Sannino, Phys. Rev. **D79**, 095008 (2009), [arXiv:0903.3115 [hep-ph]]
49. M.E. Shaposhnikov, JETP Lett. **44**, 465 (1986), [Pisma Zh. Eksp. Teor. Fiz. **44**, 364 (1986)]
50. K. Kajantie, M. Laine, K. Rummukainen, M.E. Shaposhnikov, Nucl. Phys. B **466**, 189 (1996), [arXiv:hep-lat/9510020]
51. A.A. Starobinsky, JETP Lett. **30**, 682 (1979), [Pisma Zh. Eksp. Teor. Fiz. **30**, 719 (1979)]
52. A.A. Starobinsky, Phys. Lett. B **91**, 99 (1980)
53. V.F. Mukhanov, G.V. Chibisov, JETP Lett. **33**, 532 (1981), [Pisma Zh. Eksp. Teor. Fiz. **33**, 549 (1981)]
54. A.H. Guth, Phys. Rev. **D23**, 347 (1981)
55. A.D. Linde, Phys. Lett. B **108**, 389 (1982)
56. A. Albrecht, P.J. Steinhardt, Phys. Rev. Lett. **48**, 1220 (1982)
57. F. Sannino, (2008), arXiv:0804.0182 [hep-ph]
58. P. Channuie, J.J. Joergensen, F. Sannino, JCAP **1105**, 007 (2011), [arXiv:1102.2898 [hep-ph]]
59. F. Bezrukov, P. Channuie, J.J. Joergensen, F. Sannino, (2012), arXiv:1112.4054 [hep-ph]
60. F. Sannino, Phys. Rev. **D80**, 065011 (2009), [arXiv:0907.1364 [hep-th]]
61. F. Sannino, Phys. Rev. Lett. **105**, 232002 (2010), [arXiv:1007.0254 [hep-ph]]
62. F. Sannino, Mod. Phys. Lett. A **26**, 1763 (2011), [arXiv:1102.5100 [hep-ph]]
63. M. Mojaza, M. Nardecchia, C. Pica, F. Sannino, Phys. Rev. **D83**, 065022 (2011), [arXiv:1101.1522 [hep-th]]
64. D.I. Kaiser, Phys. Rev. **D52**, 4295 (1995), [arXiv:astro-ph/9408044]
65. S. Tsujikawa, H. Yajima, Phys. Rev. **D62**, 123512 (2000), [arXiv:hep-ph/0007351]
66. F. Bezrukov, D. Gorbunov, M. Shaposhnikov, JCAP **0906**, 029 (2009), [arXiv:0812.3622 [hep-ph]]
67. A.O. Barvinsky, A.Y. Kamenshchik, A.A. Starobinsky, JCAP **0811**, 021 (2008), [arXiv:0809.2104 [hep-ph]]
68. F. Bauer, D.A. Demir, Phys. Lett. B **698**, 425 (2011) [arXiv:1012.2900 [hep-ph]]

Appendix

A.1 Basic Group Theory Relations

The Dynkin indices label the highest weight of an irreducible representation and uniquely characterise the representations. The Dynkin indices for some of the most common representations are given in Tab. A.1. For details on the concept of Dynkin indices see, for example [1, 2].

For a representation, R, with the Dynkin indices $(a_1, a_2, \ldots, a_{N-2}, a_{N-1})$ the quadratic Casimir operator reads [3]

$$
2N\, C_2(\mathrm{r}) = \sum_{m=1}^{N-1} [N(N-m)m a_m + m(N-m)a_m{}^2
$$

$$
+ \sum_{n=0}^{m-1} 2n(N-m)a_n a_m] \tag{A.1}
$$

and the dimension of R is given by

$$
d(\mathrm{r}) = \prod_{p=1}^{N-1} \left\{ \frac{1}{p!} \prod_{q=p}^{N-1} \left[\sum_{z=q-p+1}^{p} (1+a_z) \right] \right\}, \tag{A.2}
$$

which gives rise to the following structure

$$
\begin{aligned}
d(\mathrm{r}) = {} & (1+a_1)(1+a_2)\ldots(1+a_{N-1}) \times \\
& \times (1 + \tfrac{a_1+a_2}{2})\ldots(1 + \tfrac{a_{N-2}+a_{N-1}}{2}) \times \\
& \times (1 + \tfrac{a_1+a_2+a_3}{3})\ldots(1 + \tfrac{a_{N-3}+a_{N-2}+a_{N-1}}{3}) \times \\
& \times \cdots \times \\
& \times (1 + \tfrac{a_1+\cdots+a_{N-1}}{N-1}).
\end{aligned} \tag{A.3}
$$

F. Sannino, *Dynamical Stabilization of the Fermi Scale*, SpringerBriefs in Physics, DOI: 10.1007/978-3-642-33341-5, © The Author(s) 2013

Table A.1 Examples for
Dynkin indices for some
common representations

Representation	Dynkin indices
Singlet	$(000\ldots00)$
Fundamental (F)	$(100\ldots00)$
Antifundamental ($\bar{\text{F}}$)	$(000\ldots01)$
Adjoint (G)	$(100\ldots01)$
n-index symmetric (S_n)	$(n00\ldots00)$
2-index antisymmetric (A_2)	$(010\ldots00)$

The Young tableau associated to a given Dynkin index $(a_1, a_2, \ldots, a_{N-2}, a_{N-1})$ is easily constructed. The length of row i (that is the number of boxes per row) is given in terms of the Dynkin indices by the expression $r_i = \sum_i^{N-1} a_i$. The length of each column is indicated by c_k; k can assume any positive integer value. Indicating the total number of boxes associated to a given Young tableau with b one has another compact expression for $C_2(\text{r})$,

$$2NC_2(\text{r}) = N \left[bN + \sum_i r_i^2 - \sum_i c_i^2 - \frac{b^2}{N} \right], \tag{A.4}$$

and the sums run over each column and row.

A.2 Group factors and perturbative coefficients

The four-loop beta function coefficients are [4]:

$$\beta_0 = \frac{11}{3}C_2(G) - \frac{4}{3}T(r)n_f, \tag{A.5}$$

$$\beta_1 = \frac{34}{3}C_2(G)^2 - 4C_2(r)T(r)n_f - \frac{20}{3}C_2(G)T(r)n_f, \tag{A.6}$$

$$\beta_2 = \frac{2857}{54}C_2(G)^3 + 2C_2(r)^2T(r)n_f - \frac{205}{9}C_2(r)C_2(G)T(r)n_f \tag{A.7}$$
$$- \frac{1415}{27}C_2(G)^2T(r)n_f + \frac{44}{9}C_2(r)T(r)^2n_f^2 + \frac{158}{27}C_2(G)T(r)^2n_f^2,$$

$$\beta_3 = C_2(G)^4 \left(\frac{150653}{486} - \frac{44}{9}\zeta_3 \right) + C_2(G)^3T(r)n_f \left(-\frac{39143}{81} + \frac{136}{3}\zeta_3 \right) \tag{A.8}$$
$$+ C_2(G)^2C_2(r)T(r)n_f \left(\frac{7073}{243} - \frac{656}{9}\zeta_3 \right)$$
$$+ C_2(G)C_2(r)^2T(r)n_f \left(-\frac{4204}{27} + \frac{352}{9}\zeta_3 \right)$$

$$+ 46C_2(r)^3 T(r)n_f + C_2(G)^2 T(r)^2 n_f^2 \left(\frac{7930}{81} + \frac{224}{9} \zeta_3 \right)$$

$$+ C_2(r)^2 T(r)^2 n_f^2 \left(\frac{1352}{27} - \frac{704}{9} \zeta_3 \right)$$

$$+ C_2(G)C_2(r)T(r)^2 n_f^2 \left(\frac{17152}{243} + \frac{448}{9} \zeta_3 \right)$$

$$+ \frac{424}{243} C_2(G)T(r)^3 n_f^3 + \frac{1232}{243} C_2(r)T(r)^3 n_f^3$$

$$+ \frac{d_G^{abcd} d_G^{abcd}}{N_G} \left(-\frac{80}{9} + \frac{704}{3} \zeta_3 \right) + n_f \frac{d_G^{abcd} d_r^{abcd}}{N_G} \left(\frac{512}{9} - \frac{1664}{3} \zeta_3 \right)$$

$$+ n_f^2 \frac{d_r^{abcd} d_r^{abcd}}{N_G} \left(-\frac{704}{9} + \frac{512}{3} \zeta_3 \right) .$$

The coefficients of the anomalous dimension to four-loops are [5]:

$$\gamma_0 = 3C_2(r) \tag{A.9}$$

$$\gamma_1 = \frac{3}{2} C_2(r)^2 + \frac{97}{6} C_2(r)C_2(G) - \frac{10}{3} C_2(r)T(r)n_f \tag{A.10}$$

$$\gamma_2 = \frac{129}{2} C_2(r)^3 - \frac{129}{4} C_2(r)^2 C_2(G) + \frac{11413}{108} C_2(r)C_2(G)^2 \tag{A.11}$$

$$+ C_2(r)^2 T(r)n_f(-46 + 48\zeta_3) + C_2(r)C_2(G)T(r)n_f \left(-\frac{556}{27} - 48\zeta_3 \right)$$

$$- \frac{140}{27} C_2(r)T(r)^2 n_f^2$$

$$\gamma_3 = C_2(r)^4 \left(-\frac{1261}{8} - 336\zeta_3 \right) + C_2(r)^3 C_2(G) \left(\frac{15349}{12} + 316\zeta_3 \right) \tag{A.12}$$

$$+ C_2(r)^2 C_2(G)^2 \left(-\frac{34045}{36} - 152\zeta_3 + 440\zeta_5 \right)$$

$$+ C_2(r)C_2(G)^3 \left(\frac{70055}{72} + \frac{1418}{9} \zeta_3 - 440\zeta_5 \right)$$

$$+ C_2(r)^3 T(r)n_f \left(-\frac{280}{3} + 552\zeta_3 - 480\zeta_5 \right)$$

$$+ C_2(r)^2 C_2(G)T(r)n_f \left(-\frac{8819}{27} + 368\zeta_3 - 264\zeta_4 + 80\zeta_5 \right)$$

$$+ C_2(r)C_2(G)^2 T(r)n_f \left(-\frac{65459}{162} - \frac{2684}{3} \zeta_3 + 264\zeta_4 + 400\zeta_5 \right)$$

$$+ C_2(r)^2 T(r)^2 n_f^2 \left(\frac{304}{27} - 160\zeta_3 + 96\zeta_4 \right)$$

$$+ C_2(r)C_2(G)T(r)^2 n_f^2 \left(\frac{1342}{81} + 160\zeta_3 - 96\zeta_4 \right)$$

$$+ C_2(r)T(r)^3 n_f^3 \left(-\frac{664}{81} + \frac{128}{9}\zeta_3 \right)$$

$$+ \frac{d_r^{abcd} d_G^{abcd}}{N_r}(-32 + 240\zeta_3) + n_f \frac{d_r^{abcd} d_r^{abcd}}{N_r}(64 - 480\zeta_3)$$

In the above expressions ζ_x is the Riemann zeta-function evaluated at x, T_r^a with $a = 1, \ldots, N_r$ are the generators for a generic representation r with dimension N_r. The generators are normalized via $\mathrm{tr}(T_r^a T_r^b) = T(r)\delta^{ab}$ and the quadratic Casimirs are $[T_r^a T_r^a]_{ij} = C_2(r)\delta_{ij}$. The representation $r = G$ refers to the adjoint representation. The number of fermions is indicated by n_f.

The symbols d_r^{abcd} are the fourth-order group invariants expressed in terms of contractions between the following fully symmetrical tensors:

$$d_r^{abcd} = \frac{1}{6}\mathrm{Tr}\left[T_r^a T_r^b T_r^c T_r^d + T_r^a T_r^b T_r^d T_r^c + T_r^a T_r^c T_r^b T_r^d \right.$$
$$\left. + T_r^a T_r^c T_r^d T_r^b + T_r^a T_r^d T_r^b T_r^c + T_r^a T_r^d T_r^c T_r^b \right] \quad \text{(A.13)}$$

The contractions can be written purely in terms of group invariants:

$$d_r^{abcd} d_{r'}^{abcd} = I_4(r)I_4(r')d^{abcd}d^{abcd} + \frac{3N_G}{N_G + 2}T(r)T(r')$$
$$\times \left(C_2(r) - \frac{1}{6}C_2(G) \right)\left(C_2(r') - \frac{1}{6}C_2(G) \right). \quad \text{(A.14)}$$

The expressions for the relevant group invariants are given in the main text. As mentioned there, $I_4(r)$ vanished for all exceptional groups and for $SO(3)$ and $SO(4)$. The tensor d^{abcd} is representation independent, but not group independent, and the value of its contraction for the groups $SU(N)$, $SO(N)$ and $Sp(N)$ was given in [4]. Here it is only relevant to quote the $SO(N)$ case:

$$d^{abcd} d^{abcd} = \frac{N_G(N_G - 1)(N_G - 3)}{12(N_G + 2)}. \quad \text{(A.15)}$$

The relevant group factors for $SU(N)$, $SO(N)$ and $SP(2N)$ can be found in Table II of [6].

A.3 Realization of the generators for MWT

It is convenient to use the following representation of SU(4)

$$S^a = \begin{pmatrix} \mathbf{A} & \mathbf{B} \\ \mathbf{B}^\dagger & -\mathbf{A}^T \end{pmatrix}, \qquad X^i = \begin{pmatrix} \mathbf{C} & \mathbf{D} \\ \mathbf{D}^\dagger & \mathbf{C}^T \end{pmatrix}, \quad \text{(A.16)}$$

where A is hermitian, C is hermitian and traceless, $B = -B^T$ and $D = D^T$. The S are also a representation of the $SO(4)$ generators, and thus leave the vacuum invariant $S^a E + E S^{aT} = 0$. Explicitly, the generators read

$$S^a = \frac{1}{2\sqrt{2}} \begin{pmatrix} \tau^a & 0 \\ 0 & -\tau^{aT} \end{pmatrix}, \quad a = 1, \ldots, 4, \tag{A.17}$$

where $a = 1, 2, 3$ are the Pauli matrices and $\tau^4 = \mathbb{1}$. These are the generators of $SU_V(2) \times U_V(1)$.

$$S^a = \frac{1}{2\sqrt{2}} \begin{pmatrix} 0 & \mathbf{B}^a \\ \mathbf{B}^{a\dagger} & 0 \end{pmatrix}, \quad a = 5, 6, \tag{A.18}$$

with

$$B^5 = \tau^2, \quad B^6 = i\tau^2. \tag{A.19}$$

The rest of the generators which do not leave the vacuum invariant are

$$X^i = \frac{1}{2\sqrt{2}} \begin{pmatrix} \tau^i & 0 \\ 0 & \tau^{iT} \end{pmatrix}, \quad i = 1, 2, 3, \tag{A.20}$$

and

$$X^i = \frac{1}{2\sqrt{2}} \begin{pmatrix} 0 & \mathbf{D}^i \\ \mathbf{D}^{i\dagger} & 0 \end{pmatrix}, \quad i = 4, \ldots, 9, \tag{A.21}$$

with

$$D^4 = \mathbb{1}, \quad D^6 = \tau^3, \quad D^8 = \tau^1,$$
$$D^5 = i\mathbb{1}, \quad D^7 = i\tau^3, \quad D^9 = i\tau^1. \tag{A.22}$$

The generators are normalized as follows

$$\text{Tr}\left[S^a S^b\right] = \frac{1}{2}\delta^{ab}, \quad , \text{Tr}\left[X^i X^j\right] = \frac{1}{2}\delta^{ij}, \quad \text{Tr}\left[X^i S^a\right] = 0. \tag{A.23}$$

A.4 Vector Mesons as Gauge Fields

We show how to rewrite the vector meson Lagrangian in a gauge invariant way. We assume the scalar sector to transform according to a given but otherwise arbitrary representation of the flavor symmetry group G. This is a straightforward generalization of the Hidden Local Gauge symmetry idea [7, 8], used in a similar context for the BESS models [9]. At the tree approximation this approach is identical to the one introduced first in [10, 11].

Table A.2 Field content

	G	G'
M	R	1
N	$\square$	$\bar{\square}$
A_μ	1	Adj

A.4.1 Introducing Vector Mesons

Let us start with a generic flavor symmetry group G under which a scalar field M transforms globally in a given, but generic, irreducible representation R. We also introduce an algebra valued one-form $A = A^\mu dx_\mu$ taking values in a copy of the algebra of the group G, call it G', i.e.

$$A_\mu = A^a_\mu T^a, \quad \text{with} \quad T^a \in \mathscr{A}(G'). \tag{A.24}$$

At this point the full group structure is the semisimple group $G \times G'$. M does not transform under G'. Given that M and A belong to two different groups we need another field to connect the two. We henceforth introduce a new scalar field N transforming according to the fundamental of G and to the antifundamental of G'. We then upgrade A to a gauge field over G'.

The covariant derivative for N is:

$$D_\mu N = \partial_\mu N + i \tilde{g} N A_\mu. \tag{A.25}$$

We now force N to acquire the following vacuum expectation value

$$\langle N^i_j \rangle = \delta^i_j \, v', \tag{A.26}$$

which leaves the diagonal subgroup—denoted with G_V - of $G \times G'$ invariant. Clearly G_V is a copy of G. Note that it is always possible to arrange a suitable potential term for N leading to the previous pattern of symmetry breaking. v/v' is expected to be much less than one and the *unphysical* massive degrees of freedom associated to the fluctuations of N will have to be integrated out. The would-be Goldstone bosons associated to N will become the longitudinal components of the massive vector mesons.

To connect A to M we define the one-form transforming only under G via N which—in the deeply spontaneously broken phase of N—reads:

$$\frac{\text{Tr}[NN^\dagger]}{\dim(F)} P_\mu = \frac{D_\mu N N^\dagger - N D_\mu N^\dagger}{2i\tilde{g}}, \quad P_\mu \to u P_\mu u^\dagger, \tag{A.27}$$

with u being an element of G and $\dim(F)$ the dimension of the fundamental representation of G. When evaluating P_μ on the vacuum expectation value for N we recover A_μ:

$$\langle P_\mu \rangle = A_\mu. \tag{A.28}$$

At this point it is straightforward to write the Lagrangian containing N, M and A and their self-interactions. Being in the deeply broken phase of $G \times G'$ down to G_V we count N as a dimension zero field. This is consistent with the normalization for P_μ.

The simplest[1] kinetic term of the Lagrangian is:

$$L_{kinetic} = -\frac{1}{2}\mathrm{Tr}\left[F_{\mu\nu}F^{\mu\nu}\right] + \frac{1}{2}\mathrm{Tr}\left[DNDN^\dagger\right] + \frac{1}{2}\mathrm{Tr}\left[\partial M \partial M^\dagger\right]. \tag{A.29}$$

The second kinetic term will provide a mass to the vector mesons. Besides the potential terms for M and N there is another part of the Lagrangian which is of interest to us. This is the one mixing P and M. Up to dimension four and containing at most two powers of P and M this is:

$$L_{P-M} = \tilde{g}^2 r_1 \mathrm{Tr}\left[P_\mu P^\mu M M^\dagger\right] + \tilde{g}^2 r_2 \mathrm{Tr}\left[P_\mu M P^{\mu T} M^\dagger\right]$$
$$+ i\tilde{g} r_3 \mathrm{Tr}\left[P_\mu \left(M(D^\mu M)^\dagger - (D^\mu M)M^\dagger\right)\right] + \tilde{g}^2 s \mathrm{Tr}\left[P_\mu P^\mu\right]\mathrm{Tr}\left[M M^\dagger\right]. \tag{A.30}$$

The dimensionless parameters r_1, r_2, r_3, s parameterize the strength of the interactions between the composite scalars and vectors in units of $\tilde{g}$, and are therefore expected to be of order one. We have assumed M to belong to the two index symmetric representation of a generic G= SU(N). It is straightforward to generalize the previous terms to the case of an arbitrary representation R with respect to any group G. Further higher derivative interactions including N can be included systematically.

A.4.2 Further Gauging of G

In this case we add another gauge field G_μ taking values in the algebra of G. We then define the correct covariant derivatives for M and N. For N, for example, we have:

$$D_\mu N = \partial_\mu N - i g G_\mu N + i \tilde{g}, N A_\mu. \tag{A.31}$$

Evaluating the previous expression on the vacuum expectation value of N we recover the field C_μ introduced in the text. To be more precise we need to use P_μ again but with the covariant derivative for N replaced by the one in the equation above.

---

[1] Another nonminimal term is $\mathrm{Tr}\left[NFN^\dagger M (NFN^\dagger)^T M^\dagger\right]$.

Table A.3 Field content

	G	G'
M	R	1
N	□	□̄
A_μ	1	Adj
G_μ	Adj	1

A.5 The Topological Terms and Massive Spin One States

In the previous section we introduced the vector mesons as gauge bosons of a *fake* new gauge symmetry and provided a mass term resorting to an Higgsing procedure. In fact this symmetry does not exist and there is *no* notion of a *minimal way* to break it. If all the terms are included correctly one recovers, *de facto*, a non-renormalizable Lagrangian for vector mesons preserving only the correct global flavor symmetries of the problem. This, of course, is true also for the terms involving vectors, pions and the space-time $\varepsilon_{\mu\nu\rho\sigma}$ structure. This correct way to proceed was already suggested some time ago in [12]. We will review here the salient points on the analysis done in [12].

A.5.1 The ε terms for $SU(n_f) \times SU(n_f)$

We construct an effective Lagrangian which manifestly possesses the global symmetry $SU_L(n_f) \times SU_R(n_f)$ of the underlying theory. We assume that chiral symmetry is broken according to the standard pattern $SU_L(n_f) \times SU_R(n_f) \rightarrow SU_V(n_f)$. The $n_f^2 - 1$ Goldstone bosons are encoded in the $n_f \times n_f$ matrix U transforming linearly under a chiral rotation

$$U \rightarrow u_L U u_R^\dagger, \tag{A.32}$$

with $u_{L/R} \in SU_{L/R}(n_f)$. U satisfies the non linear realization constraint $UU^\dagger = 1$. We also require $\det U = 1$. In this way we avoid discussing the axial $U_A(1)$ anomaly at the effective Lagrangian level (see Ref. [10, 11, 13] for a general discussion of anomalies). We have

$$U = e^{i\frac{\Phi}{v}}, \tag{A.33}$$

with $\Phi = \sqrt{2}\Phi^a T^a$ representing the $N_f^2 - 1$ Goldstone bosons. T^a are the generators of $SU(N_f)$, with $a = 1, \ldots, N_f^2 - 1$ and $\text{Tr}\left[T^a T^b\right] = \frac{1}{2}\delta^{ab}$. v is the vacuum expectation value.

As done above we enlarge the spectrum of massive particles including vector and axial-vector fields $A^{\mu}_{L/R} = A^{\mu,a}_{L/R} T^a$.[2]

The Wess-Zumino [14] action is the first example of ε term. It can be compactly written using the language of differential forms. It is useful to introduce the Maurer-Cartan one forms:

$$\alpha = \left(\partial_{\mu} U\right) U^{-1} dx^{\mu} \equiv (dU) U^{-1}, \quad \beta = U^{-1} dU = U^{-1}\alpha U. \tag{A.34}$$

α and β are algebra valued one forms and transform, respectively, under the left and right $SU(n_f)$ flavor group. The Wess-Zumino effective action is

$$\Gamma_{WZ}[U] = C \int_{M^5} \mathrm{Tr}\left[\alpha^5\right]. \tag{A.35}$$

The price to pay in order to make the action local is to augment by one the space dimensions. Hence the integral must be performed over a five-dimensional manifold whose boundary (M^4) is the ordinary Minkowski space. The constant C is fixed to be

$$C = -i\frac{N}{240\pi^2}, \tag{A.36}$$

by comparing the current algebra prediction for the time honored process $\pi^0 \to 2\gamma$ with the amplitude predicted using Eq. (A.35) once we gauge the electromagnetic sector of the Wess-Zumino term, and N is the number of colors.

We now consider ε type terms involving the vector and axial vector particles. As for the non ε part of the Lagrangian we first gauge the WZ term under the $SU_L(n_f) \times SU_R(n_f)$ chiral symmetry group. This procedure automatically induces new ε terms [15, 16, 10, 11, 13], leading to the following Lagrangian,

$$
\begin{aligned}
\Gamma_{WZ}[U, A_L, A_R] = \Gamma_{WZ}[U] &+ 5Ci \int_{M^4} \mathrm{Tr}\left[A_L\alpha^3 + A_R\beta^3\right] \\
&- 5C \int_{M^4} \mathrm{Tr}\,[(dA_L A_L + A_L dA_L)\alpha \\
&+ (dA_R A_R + A_R dA_R)\beta] \\
&+ 5C \int_{M^4} \mathrm{Tr}\left[dA_L dU A_R U^{-1} - dA_R dU^{-1} A_L U\right] \\
&+ 5C \int_{M^4} \mathrm{Tr}\left[A_R U^{-1} A_L U \beta^2 - A_L U A_R U^{-1}\alpha^2\right] \\
&+ \frac{5C}{2} \int_{M^4} \mathrm{Tr}\left[(A_L\alpha)^2 - (A_R\beta)^2\right] \\
&+ 5Ci \int_{M^4} \mathrm{Tr}\left[A_L^3\alpha + A_R^3\beta\right]
\end{aligned}
$$

---

[2] We rescale A by the coupling constant $\tilde{g}$.

$$+ 5Ci \int_{M^4} \text{Tr} \left[(dA_R A_R + A_R dA_R) U^{-1} A_L U \right.$$

$$\left. - (dA_L A_L + A_L dA_L) U A_R U^{-1} \right]$$

$$+ 5Ci \int_{M^4} \text{Tr} \left[A_L U A_R U^{-1} A_L \alpha + A_R U^{-1} A_L U A_R \beta \right]$$

$$+ 5C \int_{M^4} \text{Tr} \left[A_R^3 U^{-1} A_L U - A_L^3 U A_R U^{-1} \right.$$

$$\left. + \frac{1}{2} (U A_R U^{-1} A_L)^2 \right]$$

$$- 5Cr \int_{M^4} \text{Tr} \left[F_L U F_R U^{-1} \right]. \tag{A.37}$$

Here the two-forms F_L and F_R are defined as $F_L = dA_L - i A_L^2$ and $F_R = dA_R - i A_R^2$ with the one form $A_{L/R} = A_{L/R}^\mu dx_\mu$. The previous Lagrangian, when identifying the vector fields with true gauge vectors, correctly saturates the underlying global anomalies.

The last term in Eq. (A.37) is a gauge covariant term which can always be added if parity is not imposed. The last term in Eq.(A.37) is not invariant under parity, so the parameter r must vanish. All the other terms are related by gauge invariance.

Imposing just global chiral invariance, together with P and C, the previous Lagrangian has ten unrelated terms [12]:

$$\Gamma_{WZ} [U, A_L, A_R] = \Gamma_{WZ} [U] + 5c_1 i \int_{M^4} \text{Tr} \left[A_L \alpha^3 + A_R \beta^3 \right]$$

$$+ 5c_2 \int_{M^4} \text{Tr} \left[(dA_L A_L + A_L dA_L) \alpha \right.$$

$$\left. + (dA_R A_R + A_R dA_R) \beta \right]$$

$$- 5c_3 \int_{M^4} \text{Tr} \left[dA_L dU A_R U^{-1} - dA_R dU^{-1} A_L U \right]$$

$$- 5c_4 \int_{M^4} \text{Tr} \left[A_R U^{-1} A_L U \beta^2 - A_L U A_R U^{-1} \alpha^2 \right]$$

$$- \frac{5c_5}{2} \int_{M^4} \text{Tr} \left[(A_L \alpha)^2 - (A_R \beta)^2 \right]$$

$$+ 5c_6 i \int_{M^4} \text{Tr} \left[A_L^3 \alpha + A_R^3 \beta \right]$$

$$+ 5c_7 i \int_{M^4} \text{Tr} \left[(dA_R A_R + A_R dA_R) U^{-1} A_L U \right.$$

$$\left. - (dA_L A_L + A_L dA_L) U A_R U^{-1} \right]$$

$$+ 5c_8 i \int_{M^4} \text{Tr} \left[A_L U A_R U^{-1} A_L \alpha + A_R U^{-1} A_L U A_R \beta \right]$$

$$- 5c_9 \int_{M^4} \text{Tr} \left[A_R^3 U^{-1} A_L U - A_L^3 U A_R U^{-1} \right]$$

$$- \frac{5c_{10}}{2} \int_{M^4} \text{Tr} \left[(U A_R U^{-1} A_L)^2 \right], \tag{A.38}$$

where the c-coefficients are imaginary. We see that while the gauging procedure of the Wess Zumino term automatically generates a large number of ε terms, it does not guarantee that we have uncovered all terms consistent with chiral, P and C invariance. Indeed there is still one new single trace term [12] to add to the action:

$$c_{11} i \int_{M^4} \text{Tr} \left[A_L^2 \left(U A_R U^{-1} \alpha - \alpha U A_R U^{-1} \right) + A_R^2 \left(U^{-1} A_L U \beta - \beta U^{-1} A_L U \right) \right], \tag{A.39}$$

and c_{11} is an imaginary coefficient. Imposing invariance under CP has been very useful to reduce the number of possible ε terms. For example it is easy to verify that a term of the type $\text{Tr} \left[d A_L \left(U A_R U^{-1} \right)^2 \right]$ is CP odd.

In Appendix A of [12] we provided a general proof that all the dimension four (i.e. 4-derivative) terms involving the Lorentz tensor $\varepsilon_{\mu\nu\rho\sigma}$, which are consistent with global chiral symmetries as well as C and P invariance, are the ones presented in Eq. (A.38) and Eq. (A.39).

A.5.2 The ε terms for $SU(2N_f)$

We consider now fermions in a pseudoreal representation, for example $SU(2)$ TC with n_f fermions in the fundamental representation. The global symmetry group is $SU(2n_f)$ and if chiral symmetry breaking occurs we expect it to break to $Sp(2n_f)$. We divide the generators T of $SU(2n_f)$, normalized according to $\text{Tr} \left[T^a T^b \right] = \frac{1}{2} \delta^{ab}$, into two classes. We call the generators of $Sp(2n_f)$ $\{S^a\}$ with $a = 1, \ldots, 2n_f^2 + n_f$, and the remaining $SU(2N_f)$ generators (parameterizing the quotient space $SU(2N_f)/Sp(2N_f)$) $\{X^i\}$ with $i = 1, \ldots, 2n_f^2 - n_f - 1$.

This breaking pattern gives $2N_f^2 - n_f - 1$ Goldstone bosons, encoded in the antisymmetric matrix U^{ij} and $i, j = 1, \ldots, 2n_f$ as follows:

$$U = e^{i \frac{\Pi^i X^i}{v}} E, \tag{A.40}$$

where the $n_f \times n_f$ matrix E is

$$E = \begin{pmatrix} 0 & 1 \\ -1 & 0 \end{pmatrix}. \tag{A.41}$$

U transforms linearly under a chiral rotation

$$U \rightarrow uUu^T, \tag{A.42}$$

with $u \in SU(2n_f)$. The non linear realization constraint, $UU^\dagger = 1$, is automatically satisfied.

The generators of the $Sp(2n_f)$ satisfy the following relation,

$$S^T E + E S = 0, \tag{A.43}$$

while the X^i generators obey,

$$X^T = E X E^T, \tag{A.44}$$

Using this last relation we can easily demonstrate that $U^T = -U$. We also require

$$Pf\, U = 1, \tag{A.45}$$

avoiding in this way to consider the explicit realization of the underlying axial anomaly at the effective Lagrangian level.

We define the following vector field

$$A_\mu = A_\mu^a T^a, \tag{A.46}$$

which formally transforms under a $SU(2N_f)$ rotation as

$$A_\mu \rightarrow u A_\mu u^\dagger - i \partial_\mu u u^\dagger. \tag{A.47}$$

We generate the ε terms following the same procedure used for the $SU_L(N_f) \times SU_R(N_f)$ global symmetry case. First we introduce the one form

$$\alpha = (dU) U^{-1}. \tag{A.48}$$

It is sufficient to define only α since the analog of $\beta = U^{-1}dU = \alpha^T$ is now not an independent form. The Wess-Zumino action term is:

$$\tilde{\Gamma}_{WZ}[U] = C \int_{M^5} Tr\left[\alpha^5\right], \tag{A.49}$$

where again we are integrating on a five dimensional manifold and $C = -i\frac{2}{240\pi^2}$ for $N = 2$. We are considering here an $SU(2)$ underlying gauge theory with fermions in the fundamental representation.

We now gauge the Wess-Zumino action under the $SU(2n_f)$ chiral symmetry group. This procedure provides single trace ε-terms involving vector, axial and

Goldstones with an universal coupling C. The gauged CP invariant Wess-Zumino term is

$$
\begin{aligned}
\tilde{\Gamma}_{WZ}[U, A] = \tilde{\Gamma}_{WZ}[U] &+ 10Ci \int_{M^4} \text{Tr}[A\alpha^3] \\
&- 10C \int_{M^4} \text{Tr}[(dAA + AdA)\alpha] \\
&- 5C \int_{M^4} \text{Tr}[dAdUA^TU^{-1} - dA^TdU^{-1}AU] \\
&- 5C \int_{M^4} \text{Tr}[UA^TU^{-1}(A\alpha^2 + \alpha^2 A)] \\
&+ 5C \int_{M^4} \text{Tr}[(A\alpha)^2] + 10C\,i \int_{M^4} \text{Tr}[A^3\alpha] \\
&+ 10C\,i \int_{M^4} \text{Tr}[(dAA + AdA)UA^TU^{-1}] \\
&- 10C\,i \int_{M^4} \text{Tr}[A\alpha AUA^TU^{-1}] \\
&+ 10C \int_{M^4} \text{Tr}\left[A^3UA^TU^{-1} + \frac{1}{4}(AUA^TU^{-1})^2\right],
\end{aligned}
\tag{A.50}
$$

where $A = A^\mu dx_\mu$. The previous Lagrangian must be generalized to be only globally invariant under a chiral rotation and invariant under CP and one obtains [12].

$$
\begin{aligned}
\tilde{\Gamma}_{WZ}[U, A] = \tilde{\Gamma}_{WZ}[U] &+ C_1 i \int_{M^4} \text{Tr}\left[A\alpha^3\right] \\
&- C_2 \int_{M^4} \text{Tr}[(dAA + AdA)\alpha] \\
&- C_3 \int_{M^4} \text{Tr}\left[dAdUA^TU^{-1} - dA^TdU^{-1}AU\right] \\
&- C_4 \int_{M^4} \text{Tr}[UA^TU^{-1}(A\alpha^2 + \alpha^2 A)] \\
&+ C_5 \int_{M^4} \text{Tr}\left[(A\alpha)^2\right] + C_6\,i \int_{M^4} \text{Tr}\left[A^3\alpha\right] \\
&+ C_7 i \int_{M^4} \text{Tr}\left[(dAA + AdA)UA^TU^{-1}\right] \\
&- C_8 i \int_{M^4} \text{Tr}\left[A\alpha AUA^TU^{-1}\right] + C_9 \int_{M^4} \text{Tr}[A^3UA^TU^{-1}] \\
&+ C_{10} \int_{M^4} \text{Tr}[(AUA^TU^{-1})^2] \\
&+ C_{11} i \int_{M^4} \text{Tr}[A^2(\alpha UA^TU^{-1} - UA^TU^{-1}\alpha)],
\end{aligned}
\tag{A.51}
$$

where C_i are imaginary. The last term is a new term not generated by gauging the Wess-Zumino effective action.

At this point the application to extensions of the SM featuring chiral dynamics is straightforward. Summarizing, the SM gauge bosons, being true gauge fields, must be introduced via the correct gauging of the Wess-Zumino term. Any other spin one field which is not a gauge degree of freedom must be introduced in the manner presented above, i.e. allowing for a very general form of the interactions with the Goldstone bosons featuring an ε tensor. Often, in literature, spin-one non-gauge degrees of freedom are introduced again as gauge degrees of freedom (see for example [17]). This latter procedure can be considered as a simple phenomenological approach.

A.6 Spectrum of Strongly Coupled Theories: Higgsless Versus Higgsful theories

Often, in the literature, a number of incorrect statements are made when discussing the spectrum of TC theories. Here we will try to clarify first the situation in QCD and then show how to use new analytic means to gain control over the spectrum of strongly coupled theories with fermions in higher dimensional representations.

One approach is based on studying the theory in the large number of colors (N) limit [18, 19]. At the same time one may obtain more information by requiring the theory to model the (almost) spontaneous breakdown of chiral symmetry [20, 21]. A standard test case, for ordinary QCD, is pion pion scattering in the energy range up to about 1 GeV. Some time ago, an attempt was made [22, 23] to implement this combined scenario. We used pion pion scattering to provide some insight on the low lying hadronic spectrum of QCD.

Before turning to the spectrum of the lightest composite states in QCD we offer a simple definition of Higgsless theory: *If the composite state with the same quantum numbers of the Higgs is not the lightest particle in the spectrum after the Goldstones then the theory is Higgsless.* In practice we will use the massive spin one states to compare the mass of the composite Higgs with.

A.6.1 The Lightest Composite Scalars in QCD

The scalar sector of QCD and any TC theory constitutes a complicated sector. For QCD, in [24], using the 't Hooft large N limit, chiral dynamics and unitarity constraints the $f_0(600)$ resonance mass was found to be around 550 MeV. Other authors [25–27] have found similar results. Such a low value would make it different from a p-wave quark-antiquark state, which is expected to be in the 1000–1400 MeV range. We assume then that it is a four quark state (glueball states are expected to be in the 1.5 GeV range from lattice investigations). Four quark states of diquark-quark type

[28, 29] and meson-meson type [30] have been discussed in the literature for many years. Accepting this picture, however, poses a problem for the accuracy of the large N inspired description of the scattering since four quark states are predicted not to exist in the large N limit of QCD. We shall take the point of view that a four quark type state is present since it allows a natural fit to the low energy data. In practice, since the parameters of the pion contact and rho exchange contributions are fixed, the sigma is the most important one for fitting and fits may even be achieved [31] if the vector meson piece is neglected. However the well established, presumably four quark type, $f_0(980)$ resonance must be included to achieve a fit in the region just around 1 GeV.

There is by now a fairly large literature on the effect of light "exotic" scalars in low energy meson meson scattering. There seems to be a consenesus, arrived at using rather different approaches (keeping however, unitarity), that the sigma exists.

Here we use two large N limits of QCD as well as our information on the low lying spectrum of QCD to extract information on the spectrum of the lightest states for strongly coupled theories with fermions in various representations of the underlying strongly coupled gauge group. Lifting the strongly coupled scale to the electroweak one for theories with underlying fermions in two index representations we will show that the light scalar with the same quantum numbers of the Higgs is lighter than the lightest techni-vector meson.

A.6.2 Scalars in the 't Hooft Large N: Higgsless Theories

We concentrate on the lightest scalar $f_0(600)$ and on the vector meson $\rho(770)$. The $q\bar{q}$ nature of the vector meson is clear. This means that its mass does not scale with the number of colors while its width decrease as $1/N$. We argued above that $f_0(600)$ is a multiquark state. In this case its mass scales with a positive power of N and its width remains constant or grows with N. In formulae:

$$m_\rho^2 \sim \Lambda_{QCD}^2, \qquad \Gamma_\rho \sim \frac{1}{N} \tag{A.52}$$

$$m_{f_0}^2 \sim N^p \Lambda_{QCD}^2, \qquad \Gamma_{f_0} \sim N^q, \tag{A.53}$$

with $p > 0$ and $q > -1$.

Scaling up these results to the electroweak theory is straightforward. We first generalize the number of technidoublets gauged under the electroweak theory as well the number of TCs N_{TC}, holding fixed the weak scale we have:

$$M_{T\rho} = \frac{\sqrt{2}v_{weak}}{F_\pi} \frac{\sqrt{3}}{\sqrt{N_D N_{TC}}} m_\rho \tag{A.54}$$

$$M_{Tf_0} = \frac{\sqrt{2}v_{weak}}{F_\pi \sqrt{N_D}} \left(\frac{N_{TC}}{\sqrt{3}}\right)^{\frac{p-1}{2}} m_{f_0}, \tag{A.55}$$

where N_D is the number of doublets, v_{weak} is the electroweak scale and the extra $\sqrt{2}$ is due to our normalization of the pion decay constant. Note that for $p = 0$ and $q = -1$ the $f_0(600)$ would scale like the ρ and would then be regarded as a quark-antiquark meson at large N. However, as we mentioned, there are, by now, strong indications that this state is not of $q\bar{q}$ nature and hence $p > 0$ and $q > -1$.

Let us choose for definitiveness $p = 1$. Already for $N_{TC} \sim 6$, for any N_D the scalar is heavier than the vector meson. Hence for fermions in the fundamental representation of the TC theory we expect *no scalars* lighter than the respective vector mesons for any N_{TC} larger than or about 6 TCs. It is hence fair to call these theories Higgsless. Note that the previous statements may be altered if the theory features walking dynamics.

A.6.3 Alternative Large N Limits

The previous results are in agreement with the common lore about the light spectrum of QCD-like theories. Interestingly even if for N = 3 one has a scalar state lighter than the lightest vector meson it becomes heavier already for N>6. Clearly the reason behind this is that, due to its multiquark nature, the lightest state possesses different scaling properties than the vector meson. The situation changes when we consider alternative extensions of QCD using higher dimensional representations. At large N different extensions capture different dynamical properties of QCD.

A.6.3.1 The Two Index Antisymmetric Fermions—Link to QCD

Consider redefining the $N = 3$ quark field with color index A (and flavor index not written) as

$$q_A = \frac{1}{2}\varepsilon_{ABC}q^{[B,C]}, \qquad q^{[B,C]} = -q^{[C,B]}, \qquad (A.56)$$

so that, for example, $q_1 = q^{23}$ and similarly for the adjoint field, $\bar{q}^1 = \bar{q}_{23}$ etc. This is just a trivial change of variables. However for $N > 3$ the resulting theory will be different since the two index antisymmetric quark representation has $N(N-1)/2$ rather than N color components. As was pointed out by Corrigan and Ramond [32], who were mainly interested in the problem of the baryons at large N, this shows that the extrapolation of QCD to higher N is not unique. Further investigation of the properties of the alternative extrapolation model introduced in [32] was carried out by Kiritsis and Papavassiliou [33].

It may be worthwhile to remark that gauge theories with two index quarks have gotten a great deal of attention. Armoni, Shifman and Veneziano [34] have proposed an interesting relation between certain sectors of the two index antisymmetric (and symmetric) theories at large number of colors and sectors of super Yang-Mills (SYM).

Fig. A.1 Two index fermion—gluon vertex

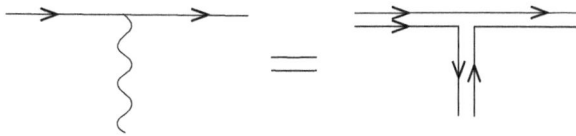

Fig. A.2 Diagram for F_π for the two index quark

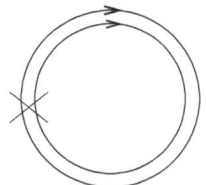

Using a supersymmetric inspired effective Lagrangian approach $1/N$ corrections were investigated in [35].

Besides these two limits a third one for massless one-flavor QCD, which is in between the 't Hooft and Corrigan Ramond ones, has been been proposed in [36]. Here one first splits the QCD Dirac fermion into the two elementary Weyl fermions and afterwards assigns one of them to transform according to a rank-two antisymmetric tensor while the other remains in the fundamental representation of the gauge group. For three colors one reproduces one-flavor QCD and for a generic number of colors the theory is chiral. The generic N is a particular case of the generalized Georgi-Glashow (gGG) model [37]. The finite temperature phase transition and its relation with chiral symmetry has been investigated in [38] while the effects of a nonzero baryon chemical potential were pioneered in [39]. More recent work in this direction has appeared in the literature [40, 41]. In particular in [41] the authors have shown that one of the high density QCD phases investigated in [39], i.e. the color superconductive one, seem to be favored at large N. This is a very interesting result which modifies and improves on the results in [39]. On the validity of the large N equivalence between different theories we refer the reader to [42, 43].

To illustrate the large N counting when quarks are designated to transform according to the two index antisymmetric representation of color SU(3) one may employ [18] the mnemonic where each tensor index of this group is represented by a directed line. Then the quark-quark gluon interaction is pictured as in Fig. A.1.

The two index quark is pictured as two lines with arrows pointing in the same direction, as opposed to the gluon which has two lines with arrows pointing in opposite directions. The coupling constant representing this vertex is taken to be $g_t/\sqrt{N}$, where g_t does not depend on N and is kept fixed.

A "one point function", like the pion decay constant, F_π has as it's simplest diagram, Fig. A.2

The X represents a pion insertion and is associated with a normalization factor for the color part of the pion's wavefunction,

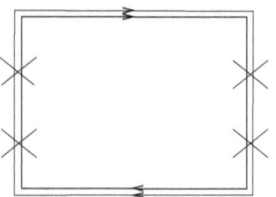

Fig. A.3 Diagram for the scattering amplitude, A with the 2 index quark

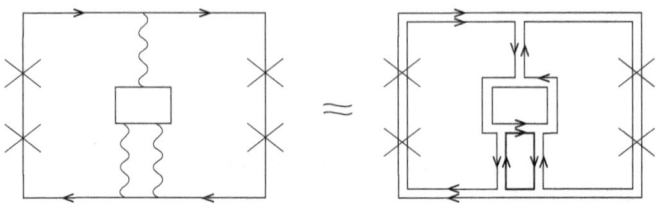

Fig. A.4 Diagram for the scattering amplitude, A including an internal 2 index quark loop

$$\frac{\sqrt{2}}{\sqrt{N(N-1)}},\tag{A.57}$$

which scales for large N as $1/N$. The two circles each carry a quark index so their factor scales as N^2 for large N; more precisely, taking the antisymmetry into account, the factor is

$$\frac{N(N-1)}{2}.\tag{A.58}$$

The product of Eqs. (A.57) and (A.58) yields the N scaling for F_π:

$$F_\pi^2(N) = \frac{N(N-1)}{6} F_\pi^2(3).\tag{A.59}$$

For large N, F_π scales proportionately to N rather than to $\sqrt{N}$ as in the case of the 't Hooft extrapolation.

Using this scaling the $\pi\pi$ scattering amplitude, A scales as,

$$A(N) = \frac{6}{N(N-1)} A(3),\tag{A.60}$$

which, for large N scales as $1/N^2$ rather than as $1/N$ in the 't Hooft extrapolation. This scaling law for large N may be verified from the mnemonic in Fig. A.3, where there is an N^2 factor from the two loops multiplied by four factors of $1/N$ from the X's.

There is still another different feature with respect to the 't Hooft expansion; consider the typical $\pi\pi$ scattering diagram with an extra internal (two index) quark loop, as shown in Fig. A.4.

In this diagram there are four X's (factor from Eq.(A.57)), five index loops (factor from Eq.(A.58)) and six gauge coupling constants. These combine to give a large N scaling behavior proportional to $1/N^2$ for the $\pi\pi$ scattering amplitude. We see that diagrams with an extra internal 2 index quark loop are not suppressed compared to the leading diagrams. This is analogous, as pointed out in [33], to the behavior of diagrams with an extra gluon loop in the 't Hooft extrapolation scheme. Now, Fig. A.4 is a diagram which can describe a sigma particle exchange. Thus in the 2 index quark scheme, "exotic" four quark resonances can appear at the leading order in addition to the usual two quark resonances. The possibility of a sigma-type state appearing at leading order means that one can construct a unitary $\pi\pi$ amplitude already at $N = 3$ in the 2 antisymmetric index scheme. From the point of view of low energy $\pi\pi$ scattering, it seems to be unavoidable to say that the 2 index scheme is more realistic than the 't Hooft scheme given the existence of a four quark type sigma.

Of course, the usual 't Hooft extrapolation has a number of other things to recommend it. These include the fact that nearly all meson resonances seem to be of the quark- antiquark type, the OZI rule predicted holds to a good approximation and baryons emerge in an elegant way as solitons in the model.

A fair statement is that each extrapolation emphasizes different aspects of $N = 3$ QCD. In particular, the usual scheme is not really a replacement for the true theory. That appears to be the meaning of the fact that the continuation to $N > 3$ is not unique.

A.6.3.2 Quarks in Two Index Symmetric Color Representation

Clearly the assignment of femions to the two index symmetric representation of color SU(3) is very similar to the previous case. We denote the fields as,

$$q_{\{AB\}} = q_{\{BA\}}. \tag{A.61}$$

There will be $N(N + 1)/2$ different color states for the two index symmetric quarks. This means that there is no value of N for which the symmetric theory can be made to correspond to true QCD. On the other hand, for large N we can make the approximation

$$A^{sym}(N) \approx A^{asym}(N), \tag{A.62}$$

for the $\pi\pi$ scattering amplitude.

As far as the large N counting goes, the mnemonics in Figs. A.1–A.4 are still applicable to the case of quarks in the two index symmetric color representation. For not so large N, the scaling factor for the pion insertion is

Fig. A.5 Diagram for meson
decay into two glueballs

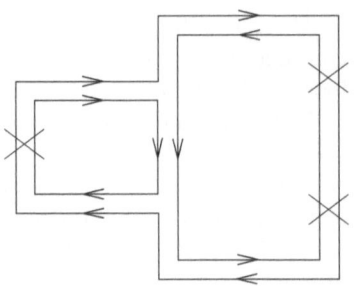

$$\frac{\sqrt{2}}{\sqrt{N(N+1)}}, \tag{A.63}$$

and the pion decay constant scales as

$$F_{\pi}^{sym}(N) \propto \sqrt{\frac{N(N+1)}{2}}. \tag{A.64}$$

With the identification $A^{QCD} = A^{asym}(3)$, the use of Eq. (A.62) enables us to estimate the large N scattering amplitude as,

$$A^{sym}(N) \approx \frac{6}{N^2} A^{QCD}. \tag{A.65}$$

In applications to minimal walking TC theories this formula is useful for making estimates involving weak gauge bosons via the Goldstone boson equivalence theorem [44].

Finally we remark on the large N scaling rules for meson and glueball masses and decays in either the two index antisymmetric or two index symmetric schemes. Both meson and glueball masses scale as $(N)^0$. Furthermore, all six reactions of the type

$$a \rightarrow b + c, \tag{A.66}$$

where a,b and c can stand for either a meson or a glueball, scale as $1/N$. This is illustrated in Fig. A.5 for the case of a meson decaying into two glueballs; note that the glueball insertion scales as $1/N$ and that two interaction vertices are involved.

A.6.4 Spectrum for Higher Dimensional Representations: Higgsful Theories

Combining our knowledge of the QCD spectrum together with the rules above for the two index antisymmetric representation we deduce the following large N scaling:

$$m_\rho^2 \sim \Lambda_{QCD}^2 \qquad \Gamma_\rho \sim \frac{2}{N(N-1)} \tag{A.67}$$

$$m_{f_0}^2 \sim \Lambda_{QCD}^2 \qquad \Gamma_{f_0} \sim \frac{2}{N(N-1)}. \tag{A.68}$$

The fact that in QCD the state $f_0(600)$ is not narrow indicates that the unknown coefficient in the expression for the width, expected to be order one, is large. However, as we increase the number of colors we expect this state to become quickly narrow. Scaling up these results for a TC theory with N_{TC} colors and fermions in the two index antisymmetric representation we have:

$$M_{T\rho} = \frac{\sqrt{2}v_{weak}}{F_\pi} \frac{\sqrt{3}\sqrt{2}}{\sqrt{N_D N_{TC}(N_{TC}-1)}} m_\rho \tag{A.69}$$

$$M_{Tf_0} = \frac{\sqrt{2}v_{weak}}{F_\pi} \frac{\sqrt{3}\sqrt{2}}{\sqrt{N_D N_{TC}(N_{TC}-1)}} m_{f_0}. \tag{A.70}$$

The input values here are the QCD masses for $f_0(600)$ and $\rho(770)$. Differently from the 't Hooft case the scalar will remain lighter than the associate technivector meson for any number of TCs. Finally, increasing the number of TCs and techniflavors we can achieve a very light scalar, lighter then its own technivector. Since in these theories one cannot differentiate a fermion-antifermion state from a multi fermion states we map the lightest scalar into the composite Higgs.

So, even without invoking walking dynamics, higher dimensional representations provide a composite Higgs lighter than the technivector meson. These theories are Higgsful for any number of colors.

One can pass from the two index antisymmetric to the two index symmetric by replacing $N_{TC} - 1$ with $N_{TC} + 1$ in the expressions above and matching the result at infinite number of colors. In Fig. A.6 the physical spectrum of spin one vector bosons and the lightest scalar is reported in TeV units in the case of two doublets ($N_D = 2$) of technifermions for different number of colors. At $N = 3$ we match the spectrum to QCD for the two index antisymmetric representation. On the left panel we draw the spectrum for the two index antisymmetric extension of QCD while on the right we consider the two index symmetric representation normalized at large N with the two index antisymmetric one. For any N_D and N_{TC} the scalar is always lighter than the associated vector meson. In the case of the two index symmetric on approaches light masses a little faster when increasing the number of colors.

Above we demonstrated that i) It is possible to have composite theories which are Higgsful ii) the resulting composite Higgs is light with respect to the TeV scale. The comparison with precision data must then be revised for these theories since the associated S parameter constraint changes. Note that in the proof we used only a straightforward geometrical scaling.

What happens to the mass of the composite Higgs in the case of walking? By increasing the number of flavors all of the composite states from the chiral-symmetric broken side become massless when reaching the fixed point since the only invariant

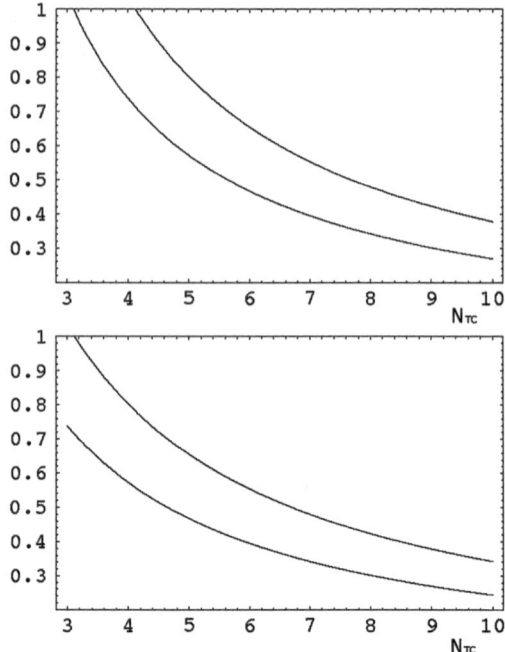

Fig. A.6 Mass of the lightest vector meson (higher curve) and scalar meson (lower curve) as function of the number of colors in TeV units. At $N = 3$ we match the spectrum to QCD for the two index antisymmetric representation. Here we use $N_D = 2$. On the left panel we draw the spectrum for the two index antisymmetric extension of QCD while on the right we consider the two index symmetric representation. Note that now for any N_D and N_{TC} the scalar is always lighter than the associated vector meson

scale of the theory vanishes there [45]. This is supported by lattice simulations [46]. We are, however, interested in the ratio between the masses of the various states to the pion decaying constant which is fixed to be the electroweak scale. Simple arguments suggest that if the transition is second order then there will be a light composite Higgs or else its mass to decay constant ratio will not vanish near the conformal point. In any event one can write a low energy effective action for the composite scalar with the quantum numbers of the Higgs—treating it as a dilaton—using trace and axial anomaly as well as chiral symmetry as done in [47]. A similar analysis using trace anomaly has been also discussed in [48]. The resulting action contains, by construction, non-analitc powers of the composite Higgs field [47] and must be treated as generating functional for the anomalous transformations of the underlying dynamics.

The possibility of a light composite Higgs in (walking) TC was first advocated in [49–52] and also proposed in [48] and [53]. Since, as shown above using standard scaling arguments, it is possible to construct TC theories with a light composite Higgs it is relevant to study its phenomenological signatures [54, 55].

References

1. E.B. Dynkin, Trans. Am. Math. Soc. **6**, 111 (1957)
2. R. Slansky, Phys. Rept. **79**, 1 (1981)
3. P. L. White, Nucl. Phys. B 403, 141 (1993) [arXiv:hep-ph/9207231].
4. T. van Ritbergen, J. A. M. Vermaseren and S. A. Larin, Phys. Lett. B 400, 379 (1997) [hep-ph/9701390].
5. J. A. M. Vermaseren, S. A. Larin and T. van Ritbergen, Phys. Lett. B 405, 327 (1997) [hep-ph/9703284].
6. C. Pica and F. Sannino, Phys. Rev. D 83, 035013 (2011) [arXiv:1011.5917 [hep-ph]].
7. M. Bando, T. Kugo, S. Uehara, K. Yamawaki, T. Yanagida, Phys. Rev. Lett. **54**, 1215 (1985)
8. M. Bando, T. Kugo, K. Yamawaki, Phys. Rept. **164**, 217 (1988)
9. R. Casalbuoni, A. Deandrea, S. De Curtis, D. Dominici, R. Gatto and M. Grazzini, Phys. Rev. D 53, 5201 (1996) [arXiv:hep-ph/9510431].
10. O. Kaymakcalan, J. Schechter, Phys. Rev. D **31**, 1109 (1985)
11. O. Kaymakcalan, S. Rajeev, J. Schechter, Phys. Rev. D **30**, 594 (1984)
12. Z. y. Duan, P. S. Rodrigues da Silva and F. Sannino, Nucl. Phys. B 592, 371 (2001) [arXiv:hep-ph/0001303].
13. P. Jain, R. Johnson, U.G. Meissner, N.W. Park, J. Schechter, Phys. Rev. D **37**, 3252 (1988)
14. J. Wess, B. Zumino, Phys. Lett. B **37**, 95 (1971)
15. E. Witten, Nucl. Phys. B **223**, 422 (1983)
16. E. Witten, Nucl. Phys. B **223**, 433 (1983)
17. K. Lane and A. Martin, arXiv:0907.3737 [hep-ph].
18. G. 't Hooft, Nucl. Phys. B 72, 461 (1974).
19. E. Witten, Nucl. Phys. B **160**, 57 (1979)
20. Y. Nambu, G. Jona-Lasinio, Phys. Rev. **122**, 345 (1961)
21. M. Gell-Mann, M. Levy, Nuovo Cim. **16**, 705 (1960)
22. F. Sannino and J. Schechter, Phys. Rev. D 52, 96 (1995) [arXiv:hep-ph/9501417].
23. M. Harada, F. Sannino and J. Schechter, Phys. Rev. D 54, 1991 (1996) [arXiv:hep-ph/9511335].
24. M. Harada, F. Sannino and J. Schechter, Phys. Rev. D 69, 034005 (2004) [arXiv:hep-ph/0309206].
25. J. R. Pelaez, Phys. Rev. Lett. 92, 102001 (2004) [arXiv:hep-ph/0309292].
26. J. A. Oller and E. Oset, Nucl. Phys. A 620, 438 (1997) [Erratum-ibid. A 652, 407 (1999)] [arXiv:hep-ph/9702314].
27. M. Uehara, arXiv:hep-ph/0308241.
28. R.L. Jaffe, Phys. Rev. D **15**, 267 (1977)
29. R.L. Jaffe, Phys. Rev. D **15**, 281 (1977)
30. J.D. Weinstein, N. Isgur, Phys. Rev. Lett. **48**, 659 (1982)
31. M. Harada, F. Sannino and J. Schechter, Phys. Rev. Lett. 78, 1603 (1997) [arXiv:hep-ph/9609428].
32. E. Corrigan, P. Ramond, Phys. Lett. B **87**, 73 (1979)
33. E.B. Kiritsis, J. Papavassiliou, Phys. Rev. D **42**, 4238 (1990)
34. A. Armoni, M. Shifman and G. Veneziano, Nucl. Phys. B 667, 170 (2003) [arXiv:hep-th/0302163].
35. F. Sannino and M. Shifman, Phys. Rev. D 69, 125004 (2004) [arXiv:hep-th/0309252].
36. T. A. Ryttov and F. Sannino, Phys. Rev. D 73, 016002 (2006) [arXiv:hep-th/0509130].
37. H. Georgi, Nucl. Phys. B **266**, 274 (1986)
38. F. Sannino, Phys. Rev. D 72, 125006 (2005) [hep-th/0507251].
39. M. T. Frandsen, C. Kouvaris and F. Sannino, Phys. Rev. D 74, 117503 (2006) [arXiv:hep-ph/0512153].
40. A. Cherman, T. D. Cohen and R. F. Lebed, Phys. Rev. D 80, 036002 (2009) [arXiv:0906.2400 [hep-ph]].
41. M. I. Buchoff, A. Cherman and T. D. Cohen, arXiv:0910.0470 [hep-ph].

42. M. Unsal and L. G. Yaffe, Phys. Rev. D 74, 105019 (2006) [arXiv:hep-th/0608180].
43. P. Kovtun, M. Unsal and L. G. Yaffe, Phys. Rev. D 72, 105006 (2005) [arXiv:hep-th/0505075].
44. B.W. Lee, C. Quigg, H.B. Thacker, Phys. Rev. D **16**, 1519 (1977)
45. R. S. Chivukula, Phys. Rev. D 55, 5238 (1997) [arXiv:hep-ph/9612267].
46. S. Catterall and F. Sannino, Phys. Rev. D 76, 034504 (2007) [arXiv:0705.1664 [hep-lat]].
47. F. Sannino and J. Schechter, Phys. Rev. D 60, 056004 (1999) [arXiv:hep-ph/9903359].
48. W. D. Goldberger, B. Grinstein and W. Skiba, Phys. Rev. Lett. 100, 111802 (2008) [arXiv:0708.1463 [hep-ph]].
49. D. K. Hong, S. D. H. Hsu and F. Sannino, Phys. Lett. B 597, 89 (2004) [arXiv:hep-ph/0406200].
50. D. D. Dietrich, F. Sannino and K. Tuominen, Phys. Rev. D 72, 055001 (2005) [arXiv:hep-ph/0505059].
51. D. D. Dietrich, F. Sannino and K. Tuominen, Phys. Rev. D 73, 037701 (2006) [arXiv:hep-ph/0510217].
52. D. D. Dietrich and F. Sannino, Phys. Rev. D 75, 085018 (2007) [arXiv:hep-ph/0611341].
53. A. Doff, A. A. Natale and P. S. Rodrigues da Silva, Phys. Rev. D 77, 075012 (2008) [arXiv:0802.1898 [hep-ph]].
54. A. R. Zerwekh, Eur. Phys. J. C 46, 791 (2006) [arXiv:hep-ph/0512261].
55. J. Fan, W. D. Goldberger, A. Ross and W. Skiba, Phys. Rev. D 79, 035017 (2009) [arXiv:0803.2040 [hep-ph]].